INSTITUTE OF GEOLOGICAL SCIENCES

Natural Environment Research Council

Mineral Assessment Report 32

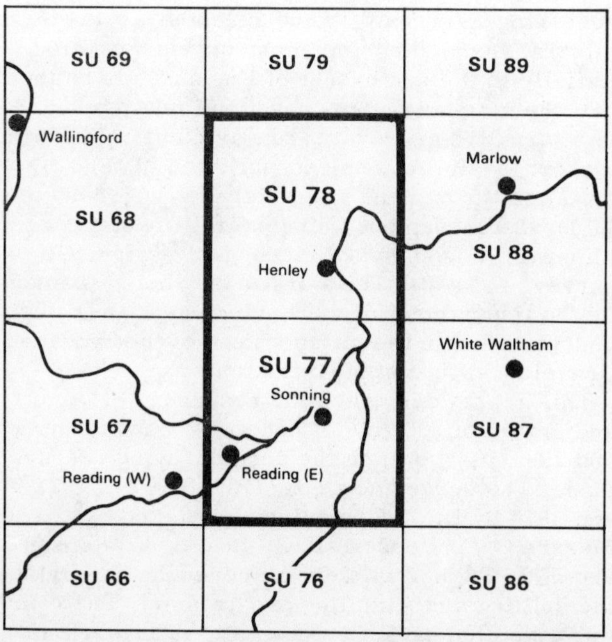

The sand and gravel resources of the country around Sonning and Henley, Berkshire, Oxfordshire and Buckinghamshire

Description of 1:25 000 resource sheets SU 77 and SU 78

H. C. Squirrell

ISBN 0 11 884032 0

London Her Majesty's Stationery Office 1978

The first twelve reports on the the assessment of British sand and gravel resources appeared in the Report Series of the Institute of Geological Sciences as a subseries. Report No. 13 onwards appear in the Mineral Assessment Report Series of the Institute. Details of published reports appear at the end of this report.

Any enquires concerning this report may be addressed to Officer-in-Charge, Industrial Minerals Assessment Unit, Institute of Geological Sciences, Keyworth, Nottingham NG12 5GQ

PREFACE

National resources of many industrial minerals may seem so large that stocktaking appears unnecessary, but the demand for minerals and for land for all purposes is intensifying and it has become increasingly clear in recent years that regional assessments of the resources of these minerals should be undertaken. The publication of information about the quantity and quality of deposits over large areas is intended to provide a comprehensive factual background against which planning decisions can be made.

Sand and gravel, considered together as naturally occurring aggregate, were selected as the bulk mineral demanding the most urgent attention, initially in the south-east of England, where about half the national output is won and very few sources of alternative aggregates are available. Following a short feasibility project, initiated in 1966 by the Ministry of Land and Natural Resources, the Mineral Assessment Unit (now the Indutrial Minerals Assessment Unit) began systematic surveys in 1968. The work is now being financed by the Department of the Environment and is being undertaken with the co-operation of the Sand and Gravel Association of Great Britain.

This report describes the resources of sand and gravel of 200 km^2 of country around Sonning and Henley, shown on the accompanying resource map. The survey was conducted between 1972 and 1974 by Dr H. C. Squirrell assisted by Messrs C. E. Corser, P. G. Hoare, P. Robson and E. J. Raynor as field officers who supervised the drilling and sampling programme. The report was compiled by H. C. Squirrell. The work is based on a 6-inch to one mile survey by J. H. Blake between 1895 and 1900, resurveyed by A. W. Kemp on sheet SU 78 in 1974-5 and amended by H. C. Squirrell on sheet SU 77 in 1975-6.

The late Mr A. P. Mace (Land Agent) was responsible for negotiating access to land for drilling. The ready co-operation of landowners and tenants in this work is appreciated. Information provided by local gravel operators, private land owners and the Berkshire and Oxfordshire County Councils is gratefully acknowledged.

A. W. Woodland
Director

Institute of Geological Sciences
Exhibition Road
London SW7 2DE

1 June 1978

SUMMARY

The geological maps of the Institute of Geological Sciences, pre-existing borehole information, and eighty-six boreholes drilled for the Industrial Minerals Assessment Unit, form the basis of the assessment of sand and gravel resources in the Sonning and Henley area of Berkshire, Oxfordshire and Buckinghamshire.

All deposits in the area which might be potentially workable for sand and gravel have been investigated and a simple statistical method has been used to estimate the volume. The reliability of the volume estimates is given at the symmetrical 95 per cent probability level.

The 1:25 000 map is divided into five resource blocks, containing between 6.5 and 16.6 km^2 of sand and gravel. For each block the geology of the deposits if described and the mineral-bearing area, the mean thicknesses of overburden and mineral and the mean gradings are stated. Detailed borehole data are also given. The geology, the boreholes and the resource blocks are shown on the accompanying map.

Bibliographical reference

SQUIRRELL, H.C. 1978. The sand and gravel resources of the country around Sonning and Henley, Berkshire, Oxfordshire and Buckinghamshire. Miner. Assess. Rep. Inst. Geol. Sci., No. 32.

Author

H.C. Squirrell, BSc, PhD.
Institute of Geological Sciences, London.

CONTENTS

FIGURES

MAP

TABLES

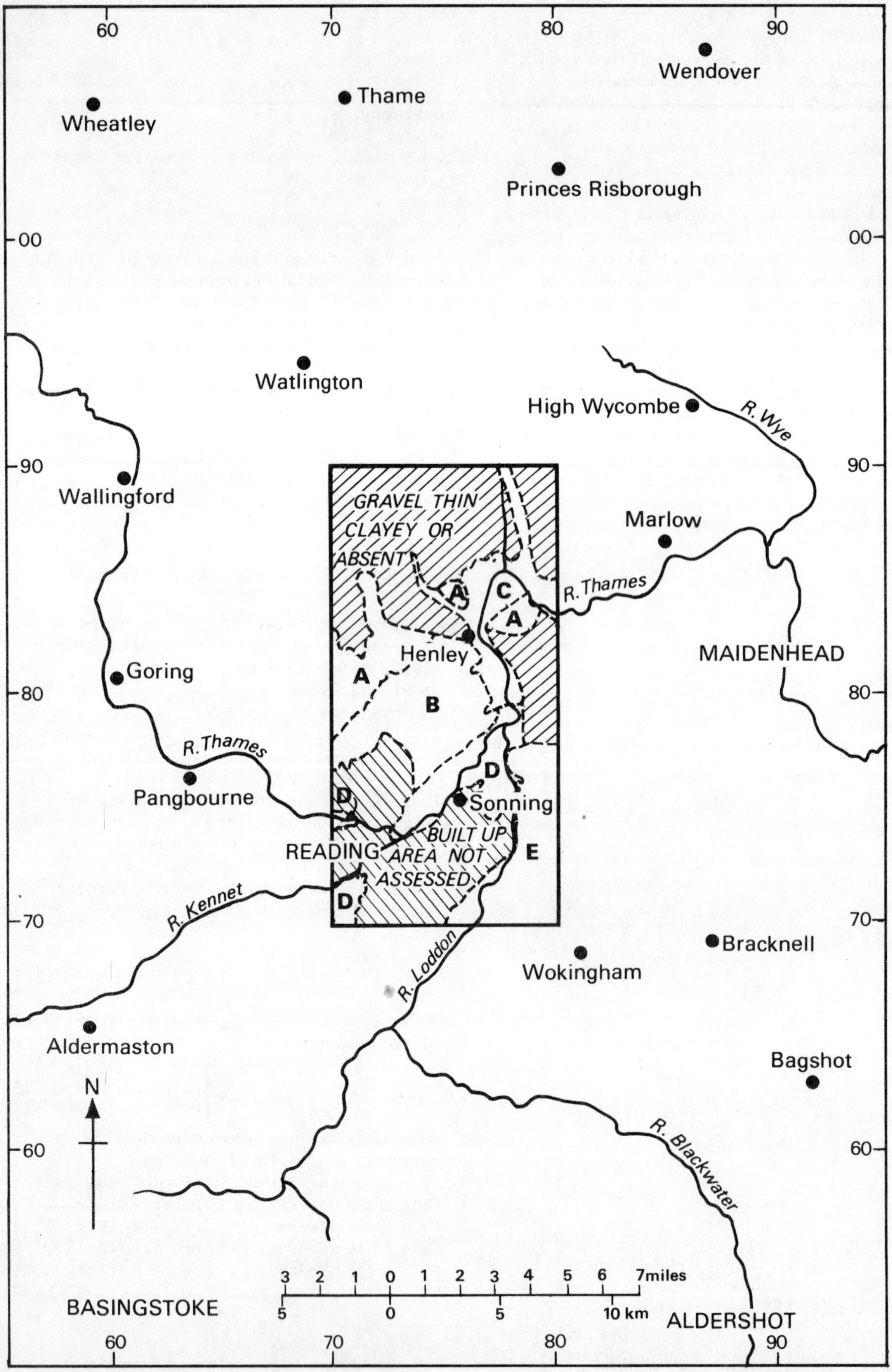

Fig. 1 Sketch map showing the location of the Sonning and Henley area and the position of the resource block boundaries

The sand and gravel resources of the country around Sonning and Henley, Berkshire, Oxfordshire and Buckinghamshire

Description of 1:25 000 resource sheets SU 77 and SU 78

H. C. Squirrell

INTRODUCTION

The survey is concerned with the estimation of resources, which include deposits that are not currently exploitable but have a foreseeable use, rather than reserves, which can be assessed only in the light of current, locally prevailing, economic considerations. Clearly, both the economic and the social factors used to decide whether a deposit may be workable in the future cannot be predicted; they are likely to change with time. Deposits not currently economically workable may be exploited as demand increases, as higher-grade or alternative materials become scarce, or as improved processing techniques are applied to them. The improved knowledge of the main physical properties of the resource and their variability which this survey seeks to provide will add significantly to the factual background against which planning policies can be decided (Archer, 1969; Thurrell, 1971; Harris and others, 1974).

The survey provides information at the 'indicated' level for which tonnage and grade are computed partly from specific measurements, samples or production data and partly from projection for a reasonable distance on geological evidence. The sites available for inspection, measurement, and sampling are too widely spaced to permit the mineral bodies to be outlined completely or the grade established throughout (Anon., 1948, p. 15).

It follows that the whereabouts of reserves must still be established, and their size and quality proved, by the customary detailed exploration and evaluation undertaken by the Industry. However, the information provided by this survey should assist in the selection of the best targets for such further work.

The following arbitrary physical criteria have been adopted: (i) the deposit should average at least one metre in thickness, (ii) the ratio of overburden to sand and gravel should be no more than 3:1, (iii) the proportion of fines (particles passing the 200-mesh BS sieve, about 1/16 mm) should not exceed 40 per cent, (iv) the deposit must lie within 25 m of the surface, this being taken as the likely maximum working depth under most circumstances. It follows from the second criterion that boreholes are drilled no deeper than 18 m if so sand and gravel has been proved.

A deposit of sand and gravel that broadly meets these criteria is regarded as 'potentially workable' and is described and assessed as 'mineral' in the report. As the assessment is at the indicated level, parts of such a deposit may not satisfy all the criteria.

For the particular needs of assessing sand and gravel resources, a grain-size classification based on the geometric scale 1/16 mm, 1/4 mm, 1 mm, 4 mm, 16 mm has been adopted. The boundaries between fines (that is, the clay and silt fractions) and sand, and between sand and gravel-grade material, are placed at 1/16 mm and 4 mm respectively (see Appendix C).

The volume and other characteristics are assessed within resource blocks, each of which, ideally, contains approximately 10 km^2 of sand and gravel. No account is taken of any factors, for example, roads, villages and high agricultural or landscape value, which might stand in the way of sand and gravel being exploited, although towns are excluded. The estimated total volume therefore bears no simple relationship to the amount that could be extracted in practice.

It must be emphasised that the assessment applies to the resource block as a whole. Valid conclusions cannot be drawn about the mineral in parts of a block, except in the immediate vicinity of the actual sample points.

DESCRIPTION OF RESOURCE SHEETS SU 77 AND SU 78

The resource sheet covers an area of 200 km^2 (77.2 square miles) of country around Sonning and Henley, in Berkshire, Oxfordshire and Buckinghamshire, of which nearly one-third is gravel-bearing. Sonning is situated a few kilometres to the north-east of Reading, in an area where sand and gravel have been worked extensively for many years. Henley lies about 10 km north-north-east of Reading, on a long stretch of the Thames Valley where the gravel remains unexploited. The mineral-bearing deposits fall into two main categories, the 'low-level' Younger River Gravels and the 'high-level' Older River Gravels. The former, (in blocks C, D and E) occupy 43.8 km^2 of the Thames and Loddon valleys and the latter, (in blocks A and B) 18.3 km^2 of the higher ground to the north and west. About 4 km^2 of the Younger River Gravels with an estimated volume of 10 million m^3 (13.1 million yd^3) have been worked out. Only 0.1 km^2 of the Older River Gravels have been worked, probably because of their high fines content (Table 2).

TOPOGRAPHY

The dominant feature is the valley of the northward-flowing River Thames, the floor of which falls gently from about 40 m (131 ft) in the south-west to 30 m (98 ft) in the north-east, a drop of only 10 m in a distance of 20 km. The valley is wider in the south and narrows northwards where it cuts deeply into the Chalk bedrock. In the south it is joined by two of its principal tributary valleys, that of the Kennet at Reading and the Loddon at Twyford (Fig. 1). In the north the valley of Hamble Brook joins that of the Thames at Mill End [785 850][1] and the dry valley running through Middle [937 857] and Lower [744 848] Assendon meets the main valley at Henley. From these valleys the ground rises northwards in a series of steps to over 183 m (600 ft) at Devil's Hill [702 892]. This higher ground is highly dissected by many minor valleys, most of which are now dry.

GEOLOGY

The northern part of the Sonning and Henley area falls within the one-inch to one mile Henley and Wallingford (254) geological sheet and the southern part within the Reading (268) sheet. For this investigation the Henley and Wallingford sheet area was resurveyed on the 1:10 000 scale by A. W. Kemp and in the Reading sheet area the existing geological lines were amended by H. C. Squirrell. The classification of both the Solid and Drift deposits is shown in Table 1.

Table 1. Classification of mapped deposits

DRIFT	Maximum recorded thickness (m)
Recent and Pleistocene	
Alluvium	4
Younger River Gravels (undifferentiated)	12.2
Head	No information
Younger Coombe Deposits	2.8
Clay-with-flints	8
Sand in Clay-with-flints	No information
Older River Gravels (undifferentiated)	16.8
Sand and gravel of unknown age	No information
SOLID	
Eocene	
Lower Bagshot Beds	5
London Clay	95
Reading Beds	27
Cretaceous	
Upper Chalk	75
Middle Chalk	48
Lower Chalk	3 (at outcrop)

Solid

The Solid deposits consist of Cretaceous and Eocene strata ranging from the uppermost few feet of the Lower Chalk through the Middle and Upper Chalk, the Reading Beds and the London Clay into the basal part of the Lower Bagshot Beds. These rocks dip very gently to the east or south-east. Only eight faults, usually small, are known in the area, four of which occur around Emmer Green [724 770]. Almost two-thirds of the area is underlain by the Chalk, mainly in the centre and north, while the Reading Beds and London Clay crop out mainly in the south. The Lower Bagshot Beds occupy only a very small area in the south-east. An account of the geology is given in the memoirs of the Henley and Wallingford, and Reading districts, respectively by Jukes-Browne and White (1908) and by Blake (1903).

Lower Chalk: The Lower Chalk crops out in a small area [730 890] in the north. Only the uppermost 2 to 3 m are present, the beds consisting of white chalk with thin grey marls, termed the *plenus* Marls, which mark the top of the formation.

Middle Chalk: The Middle Chalk is 45 to 48 m thick; it commences with the beds of the *Rhynchonella cuvieri* Zone, at the base of which lies the Melbourn Rock. This basal bed consists of 2 to 3 m of hard nodular chalk, which commonly forms a conspicuous topographical feature. It is overlain by about 15 m of softer chalk containing only scattered nodules of harder chalk. The remainder of the Middle Chalk falls within the *Terebratulina* Zone, which consists of 30 to 43 m of thick beds of homogeneous, fossiliferous, white chalk with thin beds of soft, grey, shaly marl occurring at intervals ranging from only a few metres to as much as 12 m. Flint nodules occur in only a few of the beds.

Upper Chalk: The Upper Chalk is 65 to 75 m thick. At its base, the Chalk Rock consists of cream-coloured, fossiliferous limestone containing glauconite grains and hard, green-coated nodules of slightly phosphatic chalk; it is up to one metre thick. The basal bed is overlain by soft, irregularly bedded, white chalk, which characteristically contains numerous courses of nodular brown to black flint. The chalk is fossiliferous and, on the basis of the echinoids, the zones of *Holaster planus*, *Micraster cortestudinarium* and *Micraster coranguinum* have been recognised.

Reading Beds: Lying unconformably on the Upper Chalk, the Reading Beds consist of 21 to 27 m of variably coloured clays and sands. The sequence usually commences with the Bottom Bed, 1.8 to 3 m of stiff, dark bluish grey clay interstratified with brown and olive-green glauconitic sands. It is overlain by up to 12 m of white, buff or brown sands interbedded with thin clays; locally the sands are absent. The sequence is completed by 9 to 15 m of variegated clays, most commonly grey, green, red, brown and orange, which contain only thin beds of sand.

London Clay: Resting conformably on the Reading Beds, the London Clay commences with a persistent 'basement-bed', 2 to 5 m thick, which consists mainly of fossiliferous, interstratified clays and brown and olive-green glauconitic sands and sandstone. Locally, it contains septaria, flint pebbles, lignite and iron pyrite. The 'basement bed' is overlain by 75 to 90 m of stiff, brown-weathering, bluish grey clay with beds of septaria,

1 National Grid References in this report all lie within the 100-km square SU.

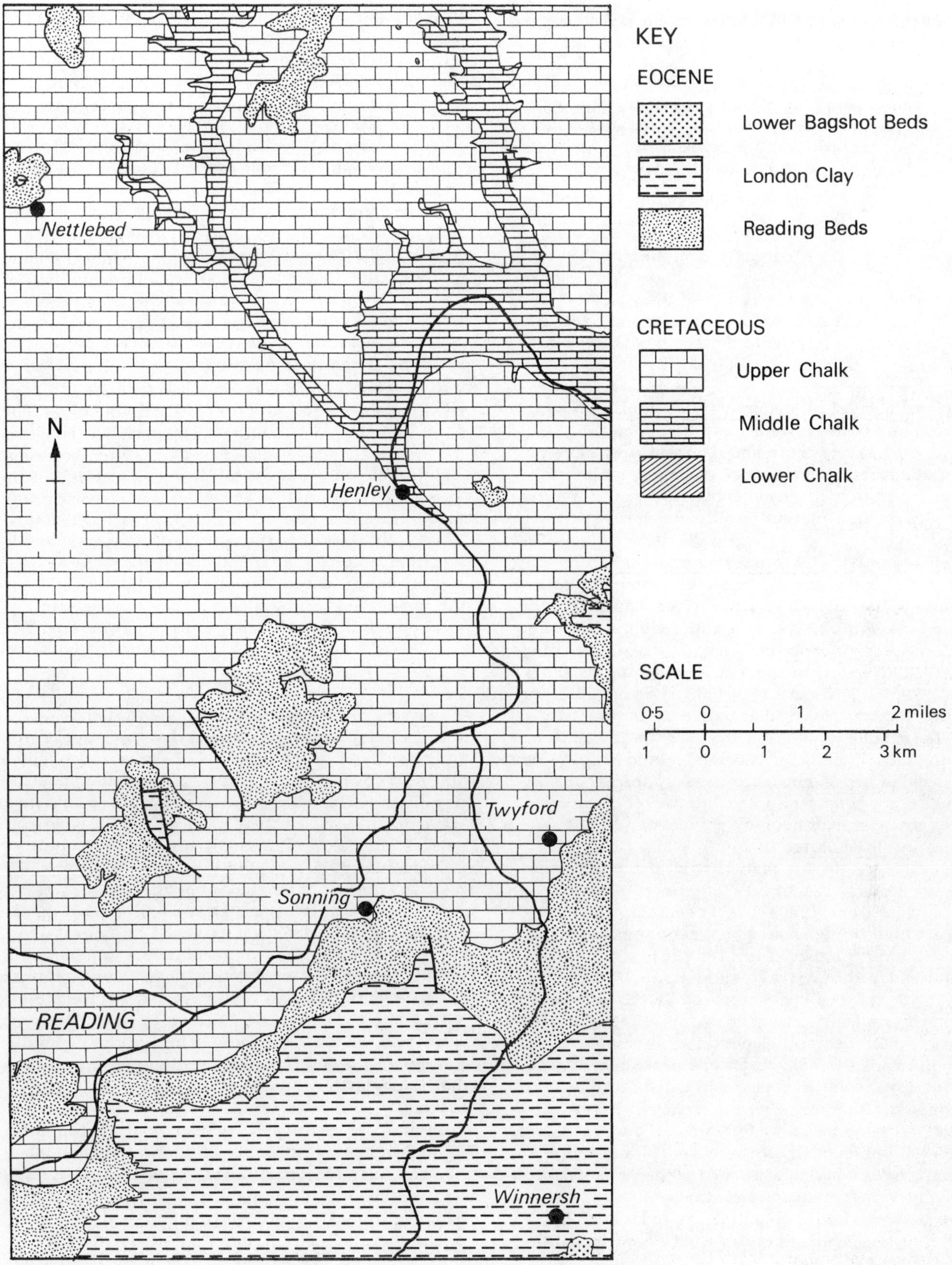

Fig. 2 Solid geology

which pass up into sandy clays at the top of the formation.

Lower Bagshot Beds: Only the lowest few metres of the Lower Bagshot Beds crop out in this area, at Wokingham Within [794 701] in the south-east. The beds consist of buff, brown, yellow, grey and white, commonly current-bedded sands with thin beds of clay.

Drift

The extensive Drift deposits contain large amounts of potentially workable sands and gravels, which occur mainly in the Older River Gravels and in the Younger River Gravels. Present-day supplies are obtained mainly from the latter, which are the cleaner and more gravelly of the two.

Sand and Gravel of unknown age: small patches of Sand and Gravel of unknown age occur in the north; because of their small extent they have not been assessed. The deposits consist mainly of subrounded to well rounded quartz, quartzite and flint gravel in a matrix of clayey sand.

Older River Gravels (undifferentiated): Formerly called Plateau Gravel, the Older River Gravels comprise the terraces of the River Thames and its tributaries that lie topographically above the more recently deposited Younger River Gravels. They lie at several different levels ranging from about 54 m (177 ft) to 155 m (508 ft) above OD, but in this report, to avoid anticipating or prejudicing the results of current work of the Institute of Geological Sciences on the Henley and Wallingford (254) Sheet and surrounding areas, their correlation has not been attempted. However, a broad differentiation has been made between 'higher' and 'lower' deposits, in Block A and Block B respectively. In the central part of the area the two are clearly demarcated by a conspicuous south-west to north-east terrace back-slope, along which the boundary between them has been drawn, from [700 783] to [746 826].

The Older River Gravels are up to 16.8 m in thickness and are characterised by the presence, in their gravel fraction, of material from distant sources, for example, quartzite, vein quartz and igneous rocks (p. 6), that were probably largely derived from the Bunter Pebble Beds of the Midlands. In contrast to the Younger River Gravels they contain a higher mean percentage of fines (20 per cent in blocks A and B) and their mean gravel content of around 50 per cent (Table 2) is about 10 per cent lower. Consequently, they have less economic potential, as evidenced by the very few and small areas of worked-out ground on their outcrop.

Clay-with-flints and Sand in Clay-with-flints: Clay-with-flints occurs extensively in the north, on the outcrop of the Chalk. It usually consists of reddish brown to dark brown, slightly sandy and silty clay containing mainly unworn nodules of flint, with some well rounded fine pebble- to cobble-sized flint, quartz and quartzite. The deposit is usually only about 2 to 3 m in thickness, but, locally, as much as 8 m have been recorded. In some parts it contains thin beds of sand, which have been identified separately as Sand in Clay-with-flints to the south-east of Highmoor [701 848] and west of Maidens Grove [722 888].

Younger Coombe Deposits: The Younger Coombe Deposits are valley-fill or fan-like accumulations that have resulted from the solifluction of loose debris lying on the Chalk surface. They are usually composed mainly of unworn flint nodules in a matrix of fragmentary or powdered chalk. Minor constituents include pebbles (of quartzite, quartz, flint, igneous rocks, etc.) and silty clay derived from other superficial deposits, mainly the Older River Gravels. Locally, they are sufficiently gravelly to constitute a mineral deposit, for example, in the top of borehole 78 SE 17, where about 2.8 m of Younger Coombe Deposits, which grade out as 'mineral', pass into the underlying Younger River Gravels.

Head: Two small areas of Head have been mapped at Sonning Common [705 800]. The deposits consist of a mixture of chalk, clay, silt, sand and gravel, in widely varying proportions, derived by solifluction of the adjacent Older River Gravels and probably also of the upper surface of the underlying Upper Chalk. Head is usually neither thick enough, nor sufficiently extensive, to form a deposit of potential economic significance.

Younger River Gravels (undifferentiated): Extensive spreads of Younger River Gravels lie in the valleys of the Thames, Kennet and Loddon. They are overlain by Alluvium along the present-day flood plain. Ranging up to 12.2 m in thickness, the deposits constitute a good source of sand and gravel (blocks C, D and E), though their quality is variable (p. 6). Their mean gravel percentage is commonly around 60 per cent and their mean fines content (mainly silt with only a small amount of clay) does not exceed 12 per cent (Table 2). Up to four terrace levels can be recognised, but in this report they have been designated 'undifferentiated', as they have not yet been classified into the standard Thames Valley terrace sequence. Details of the grading of the deposits are given in the block descriptions (pp. 8 to 13) and information about their composition is given below.

Alluvium: Forming the present-day flood plains, Alluvium consists mainly of brown, or greyish brown, silty clay containing scattered pebbles and thin beds of peat; it is up to 4 m in thickness. Abundant small lamellibranch and gastropod shells occur locally. The Alluvium usually overlies terrace sands and gravels, except locally in the main valleys, and in the small tributary valleys, where it rests on bedrock.

COMPOSITION OF THE SAND AND GRAVEL DEPOSITS

Younger River Gravels

The Younger River Gravels of the Thames, Kennet and Loddon (Fig. 1) and their minor

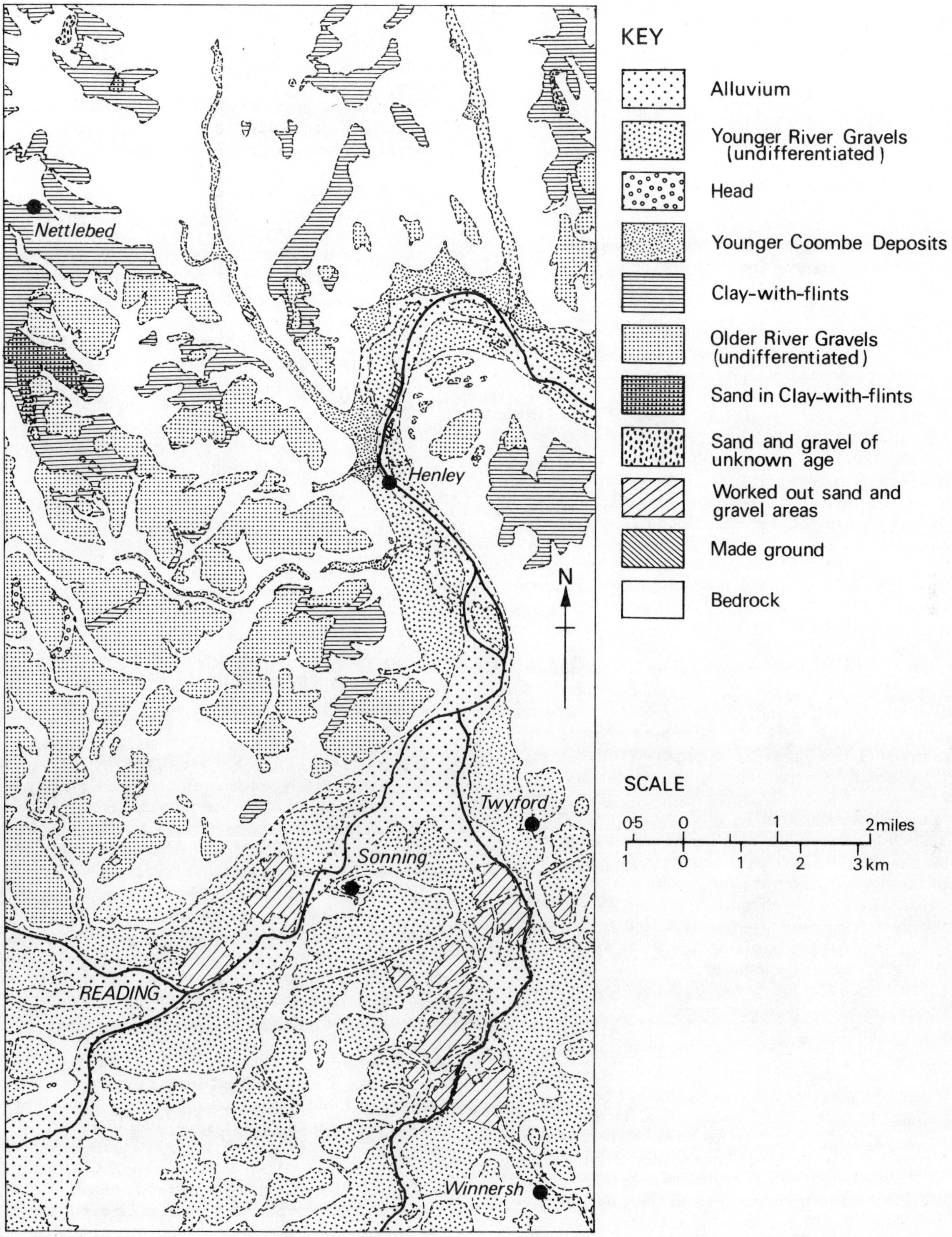

Fig. 3 Drift geology

tributaries lie within blocks C, D and E. The gravel fraction accounts for a mean of 59 per cent in Block C, 62 per cent in Block D and 57 per cent in Block E. In the Loddon and Kennet valleys the gravel consists dominantly of angular to rounded flint (over 90 per cent) with only small amounts of rounded vein quartz, quartzite and sandstone. The gravels of the Thames Valley consist mainly of subangular to rounded flint which, in the 4.75 to 9.5 mm fraction, makes up between 47 and 85 per cent of the deposit. Limestone is the second most common constituent; in the southern part of the valley around Reading and north of Sonning it comprises between 15 and 32 per cent of the 4.75 to 9.5 mm fraction and around Henley and further downstream 10 per cent or less. Other constituents include vein quartz, chalk and ironstone, none of which makes up more than 10 per cent of the deposit.

The sand makes up 30, 32 and 31 per cent of the deposit in blocks C, D and E respectively. In the Kennet and Loddon valleys it is composed almost entirely of quartz and flint, with the latter being more common in the coarser fraction. In the Thames Valley quartz and flint are the dominant constituents with limestone fragments and ooliths occurring in small quantities and, locally, chalk. The fines fraction, which accounts for 11, 6 and 12 per cent in blocks C, D and E respectively, is usually composed of silt with only a small amount of clay.

The relatively low fines content of the mineral in the Younger River Gravels is reflected in their overall classification; out of the 37 boreholes that drilled through mineral, 24 proved gravel, 9 proved 'clayey' gravel, one proved 'clayey' sandy gravel and 3 proved 'very clayey' gravel.

Older River Gravels

The Older River Gravels lie within resource blocks A and B. Gravel accounts for about half of the deposit, 53 per cent in Block A and 47 per cent in Block B, around 10 per cent less than in the Younger River Gravels (Table 2). The gravel consists dominantly of angular to well rounded flint (usually 80 per cent or more) with subrounded to well rounded quartzite, vein quartz, igneous rocks, sandstone and ironstone. The sand, which makes up about one third of the deposit, is mainly of subangular to subrounded quartz and flint, with the latter being predominant in the coarser part. The fines fraction accounts for 20 per cent of the deposit in both blocks, which is about twice as much as in the Younger River Gravels of blocks C and E and over three times as much as in those of Block D. The fines fraction generally contains more clay and less silt than in the Younger River Gravels, and where the clay content is locally particularly high the Older River Gravels have the characteristics of a 'hoggin'. The higher fines content is reflected in the grain-size classification of the Older River Gravels; 11 of the 22 Industrial Minerals Assessment Unit boreholes that found mineral proved 'clayey' gravel and 11 proved 'very clayey' gravel.

THE MAP

The sand and gravel resource map is folded into the pocket at the end of this report. The base map is the Ordnance Survey 1:25 000 Outline Edition in grey, on which the topography is shown by contours in green, the geological data in black and the mineral resource information in shades of red.

Geological data

The geological boundary lines are taken from the geological maps of the area which were surveyed on the scale of either 1:10 560 (on SU 77) or 1:10 000 (on SU 78). Borehole data, which include the stratigraphic relations and mean particle size distribution of the sand and gravel samples collected during the assessment survey, are also shown. The geological boundaries are the best interpretations of the information available at the time of survey. However, it is inevitable that local irregularities or discrepancies will be revealed by some boreholes. These are taken into account in the assessment of resources (see below and Appendix B).

Mineral resource information

The mineral-bearing ground is divided into resource blocks (see Appendix A). Within a resource block the mineral may be subdivided into areas where it is 'exposed' and areas where it is present beneath overburden. The mineral is identified as 'exposed' where the overburden, commonly consisting only of soil and subsoil, averages less than 1 m (3.5 ft) in thickness, although in some areas, for example, near present streams, the overburden, including Alluvium, may be thicker locally.

Beneath overburden the mineral may be continuous or discontinuous. As potentially workable sand and gravel was proved in all of the boreholes drilled through overburden, the mineral is regarded as continuous.

Areas where bedrock crops out and where sand and gravel does not satisfy the definition of 'mineral' are uncoloured on the map. In such areas it has been assumed that mineral is absent except in infrequent and relatively minor patches which can neither be outlined nor assessed quantitatively in the context of this survey. Areas of unassessed sand and gravel are indicated by a red stipple.

The area of the exposed sand and gravel is measured from the mapped geological boundary lines. The whole of this area is considered as mineral, although it may include small areas where sand and gravel is not present or is not potentially workable. An inferred boundary (for which a distinctive zigzag symbol is used) has been inserted to distinguish between areas of exposed mineral and areas where sand and gravel is interpreted to be not potentially workable, or absent. The zigzag symbol is intended to convey an approximate location within a likely zone of occurrence rather than to represent the breadth of the zone, its size being limited only by cartographic considerations. For the purpose of measuring areas, the centre-line of the symbol is used.

Table 2 Summary of statistical results

Resource block	Area		Mean thickness				Volume of mineral				Mean grading percentages		
	Block	Mineral	Overburden		Mineral				Limits at the 95 per cent confidence level		Fines	Sand	Gravel
	km^2	km^2	m	ft	m	ft	million m^3	million yd^3	± %	± Vol. million m^3	−1/16 mm %	−4+1/16 mm %	+4 mm %
A	15.1	6.5	0.8	2.5	6.7	22.0	44	58	40	18	20	27	53
B	26.6	11.8	0.7	2.5	4.7	15.5	55	72	20	11	20	33	47
C	14.5	13.2	1.3	4.5	6.4	21.0	84	110	24	20	11	30	59
D	18.0	16.6	1.7	5.5	5.0	16.5	83	109	24	20	6	32	62
E	20.0	14.0	1.0	3.5	2.4	8.0	34	44	24	8	12	31	57
A to E	94.2	62.1	1.2	4.0	4.9	16.0	304	398	13	40			

RESULTS

The statistical results are summarised in Table 2. Fuller particulars of particle size distributions are shown in Fig. 4. The block boundaries have been drawn to separate the areas of Older River Gravels (blocks A and B) from the areas of Younger River Gravels (blocks C, D and E). The Older River Gravels of Block A lie at a higher elevation than those of Block B and they are more variable in thickness as shown by the confidence limits of 40 and 20 per cent for the respective blocks (Table 2). Parts of the Older River Gravels have been excluded from the statistically assessed areas either because their physical characteristics do not classify them within the defined limits (p. 1) of a mineral deposit, or because they occur in outcrops of only limited extent. The Younger River Gravels of blocks C and D include mainly sands and gravels of the Thames Valley and to a small extent of the Kennet Valley; the boundary between the two blocks has been drawn arbitrarily. The deposits of Block E lie in the Loddon Valley. The compositional variations of the deposits of the Thames, Kennet and Lodden are described on pp. 4 to 6.

Accuracy of results

For the five resource blocks (A to E) the limits of error of the results of the statistical assessment at the symmetrical 95 per cent probability level (that is, it is probable that nineteen times out of twenty the true volume lies within the given limits) vary between 20 and 40 per cent. However, the true values are more likely to be nearer the volume calculated than either of the limits. Moreover, it is probable that roughly the same percentage limits would apply for the estimate of volume of a much smaller parcel of ground (say 100 hectares) containing similar sand and gravel deposits if the results from the same number of sample points (as provided by, say, ten boreholes) were used in the calculation. Thus, if closer limits are needed for the quotation of reserves of parts of a block, it can be expected that data from more than ten sample points will be required, even if the area is quite small. This point can be illustrated by considering the whole of the statistically assessed sand and gravel on the sheet. The volume, 304 million m^3, can be estimated to limits of $\pm$ 13 per cent at the symmetrical 95 per cent probability level, by a calculation based on 75 data points in blocks A to E.

However, it must be emphasised that the quoted volumes of sand and gravel bear no simple relationship to the amount that could be extracted in practice, as no allowance has been made in the calculations for any restraints (such as existing buildings and roads) on the use of land for mineral working.

NOTES ON RESOURCE BLOCKS

Block A

This block, which is divided into three parts, contains 6.5 m^2 of mineral-bearing Older River Gravels lying at between 79 m (260 ft) and 130 m (427 ft) above OD. The eastern part contains 0.6 km^2 of mineral, the central part 0.3 km^2 and the western part 5.6 km^2. The drift deposits are underlain mainly by the Upper Chalk and to a small extent by the Reading Beds. Sand and gravel has been worked only on a small scale for local use.

The assessment is based on 9 Industrial Minerals Assessment Unit (IMAU) boreholes. The mineral ranges in thickness from 2.1 m in Borehole 78 SW 13 to 12.1 m in Borehole 78 SW 14 and has a mean thickness of 6.7 m. The estimated volume of mineral is 44 million m^3 $\pm$ 40 per cent. The overburden, mainly soil and subsoil, is generally not greater than 0.3 m in thickness, but as much as 1.8 m and 4.0 m were recorded in Boreholes 78 SW 13 and 78 SW 14 respectively; the mean thickness of the overburden is 0.8 m.

The fines content (clay and silt) of the mineral ranges between 14 per cent in Borehole 78 SW 14 and 27 per cent in Borehole 78 SW 5. The proportion of sand, which is dominantly medium-grained, usually ranges between 20 and 30 per cent, but reaches 37 per cent in Borehole 78 SW 14. The mainly flint and quartz gravel makes up about half of the mineral deposit; the range is from 49 to 60 per cent. The mean grading for the block is fines 20 per cent, sand 27 per cent, gravel 53 per cent; overall, the mineral is classified as 'very clayey' gravel.

Block B

This block contains 11.8 km^2 of mineral-bearing Older River Gravels lying at between 54 m (177 ft) and 93 m (305 ft) above OD. They are underlain mainly by the Upper Chalk and, in the central part of the block, by the Reading Beds. Sand and gravel has been worked only on a small scale; the largest working [7125 7770] is only 0.05 km^2 in area.

The assessment is based on 13 IMAU boreholes and one Hydrogeological Department record. The mineral varies in thickness from 2.4 m in Borehole 78 SW 19 to 7.1 m in Borehole 77 NW 55; the mean is 4.7 m. The estimated volume of mineral is 55 million m^3 $\pm$ 20 per cent. The overburden, mainly gravelly soil and subsoil, is usually less than 0.5 m in thickness, but locally increases to over 1.0 m and reaches a maximum recorded thickness of 2.8 m in Borehole 78 SE 15. The mean is 0.7 m.

The fines content (clay and silt) of the mineral ranges between 12 and 27 per cent, except in Borehole 77 NE 17, where it reaches 35 per cent. The volume of sand, which is mainly medium-grained, commonly varies between 22 and 37 per cent, but increases to 42 per cent in Borehole 77 NW 5, and to as much as 64 per cent in Borehole 77 NW 58. The proportion of gravel usually exceeds 50 per cent and reaches a maximum of 63 per cent in Borehole 77 NW 57, but is as low as 31 per cent in Borehole 77 NW 51 and only 12 per cent in Borehole 77 NW 58. The mean grading for the block is fines 20 per cent, sand 33 per cent, gravel 47 per cent; overall, the mineral is classified as 'very clayey' gravel.

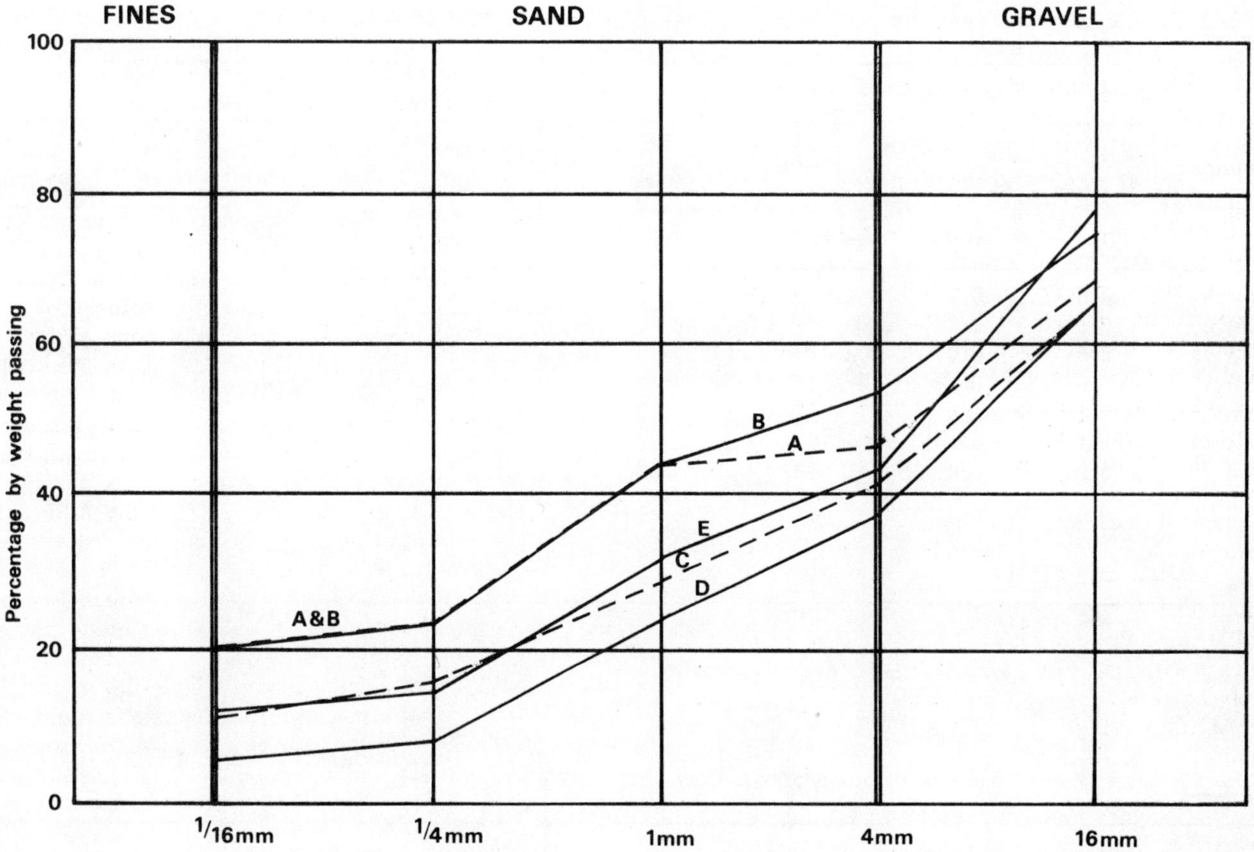

Resource Block	Percentage by weight passing				
	16 mm	$\frac{1}{4}$ mm	1 mm	4 mm	16 mm
A	20	23	44	47	69
B	20	23	44	53	75
C	11	16	29	41	66
D	6	8	24	38	66
E	12	15	32	43	78

Fig. 4 Particle-size distribution for the assessed thickness of sand and gravel in resource blocks A and E

Table 3 Data from assessment boreholes: Block A

Borehole Number	Recorded thickness: Mineral (m)	Overburden (m)	Mean grading percentages: Fines −1/16 mm	Fine sand $-\frac{1}{4}+1/16$ mm	Medium sand $-1+\frac{1}{4}$ mm	Coarse sand −4+1 mm	Fine gravel −16+4 mm	Coarse gravel +16 mm
78 SW 5	8.8	0.1	27	3	9	8	23	30
SW 6	5.9	0.3	19	2	21	9	20	29
SW 7	3.8	0.3	20	4	19	5	18	34
SW 10	7.6	0.1	23	2	17	7	18	33
SW 13	2.1	1.8	25	3	12	5	16	39
SW 14	12.1	4.0	14	4	25	8	22	27
SW 16	3.0	0.2	21	2	19	6	25	27
78 SE 14	9.5	0.2	17	1	12	8	25	35
SE 22	7.7	0.2	23	2	18	5	26	32
Mean for block			20	3	21	3	22	31

Table 4 Data from assessment boreholes: Block B

Borehole Number	Recorded thickness: Mineral (m)	Overburden (m)	Mean grading percentages: Fines −1/16 mm	Fine sand $-\frac{1}{4}+1/16$ mm	Medium sand $-1+\frac{1}{4}$ mm	Coarse sand −4+1 mm	Fine gravel −16+4 mm	Coarse gravel +16 mm
77 NW 51	3.0	0.2	27	5	30	7	16	15
NW 52	5.9	0.3	20	1	20	7	15	37
NW 53	4.5	0.2	13	3	25	9	30	20
NW 54	6.2	0.1	17	4	20	11	30	18
NW 55	7.1	0.2	18	2	20	10	23	27
NW 56	4.7	0.1	18	2	23	11	27	19
NW 57	2.7	1.2	12	3	9	13	33	30
NW 58	5.2	2.3	24	8	47	9	9	3
77 NE 17	4.7	0.4	35	4	11	7	9	34
NE 18	4.7	1.1	15	0	21	11	28	25
78 SW 18	6.5	0.2	19	1	12	9	23	36
SW 19	2.4	0.4	24	8	12	7	17	32
78 SE 15	4.1	2.8	16	3	21	9	30	21
Mean for block			20	3	21	9	22	25

Block C

Block C includes 13.2 km^2 of mineral-bearing Younger River Gravels with some Younger Coombe Deposits (see p. 4) that continuously floor the Valleys of the Thames and its tributaries around Henley. Over most of the area the deposits lie at between 20 m (66 ft) and 37 m (121 ft) above OD, and in the valley of Hamble Brook, which runs northwards to the edge of the sheet, they rise to 65 m (213 ft) above OD. About one-third of the mineral is covered by Alluvium and a small area by Younger Coombe Deposits. The superficial deposits are underlain by the Middle and Upper Chalk. Sand and gravel has been worked only on a small scale, for local uses.

The assessment is based on 15 IMAU boreholes and 7 Hydrogeological Department records. The mineral varies in thickness from 1.8 m in Borehole 254/435 to 10.8 m in Borehole 78 SE 19: the mean is 6.4 m. The estimated volume of mineral is 84 million $m^3 \pm 24$ per cent.

The overburden, usually soil overlying alluvial clay and silt with a trace of gravel, has an average thickness of 1.3 m; it ranges from a thin soil cover only to 3.6 m of silty clay in Borehole 78 SE 18. Where Alluvium has been mapped on the Younger River Gravels the overburden has a mean thickness of 2.8 m and in the Alluvium-free areas it has a mean thickness of 0.8 m.

The fines content (mainly silt) is as low as only 2 per cent in Boreholes 78 SE 18 and 27 and is as high as 29 per cent in Borehole 77 NE 22, but most commonly it ranges between 9 and 18 per cent. The proportion of sand, which is mainly medium- and coarse-grained, usually ranges between 18 and 32 per cent, but reaches 40 per cent in Borehole 78 SE 21 and 45 per cent in Borehole 78 SE 23. The proportion of gravel is commonly over 55 per cent and reaches a recorded maximum of 74 per cent in Borehole 78 NE 11. It falls to 39 and 37 per cent in Boreholes 77 NE 22 and 78 SE 23 respectively. The mean grading for the block is fines 11 per cent, sand 30 per cent, gravel 59 per cent; overall, the mineral is classified as 'clayey' gravel.

Block D

This block, which contains 16.6 km^2 of mineral-bearing Younger River Gravels, is divided into three parts separated by the unassessed built-up area of Reading. The largest area, near Sonning, includes 12.4 km^2 of mineral, the south-western area 3.6 km^2 and the smallest area only 0.6 km^2. The deposits lie between 25 m (82 ft) and 55 m (180 ft) above OD and have a flat or only gently sloping upper surface. They are underlain mainly by the Upper Chalk and to a small extent by the Reading Beds in the south-west. At the time of the survey sand and gravel was being worked at Dean's Farm, Caversham [7340 7440] and at Sonning Eye [7480 7550].

The assessment is based on 13 IMAU boreholes, 3 Hydrogeological Department Records and 140 other boreholes. The mineral varies in thickness from under 1 m locally to 10.5 m in Borehole 268/427; it has a mean of 5 m. The estimated volume of mineral is 83 million $m^3 \pm 24$ per cent. The overburden, usually soil overlying alluvial clay and silt with traces of gravel, has an average thickness of 1.7 m; it ranges from a thin soil cover only to 3.8 m in Borehole 77 SW 8. Where Alluvium has been mapped on the Younger River Gravels the overburden has a mean thickness of 2 m. Elsewhere, the mean thickness is 1.2 m.

The fines content (mainly silt) of the mineral is generally less than 10 per cent; a greater figure was proved in only one Borehole, 77 NE 25, where 17 per cent was recorded. The proportion of sand, which is dominantly medium- and coarse-grained, usually ranges between 24 and 36 per cent, but drops to 20 per cent in Borehole 77 NE 25 and reaches 43 per cent in Borehole 77 SW 14. The gravel content is over 50 per cent in every IMAU borehole and reaches a maximum of 73 per cent in Borehole 77 SW 11. The mean grading for the block is fines 6 per cent, sand 32 per cent, gravel 62 per cent; overall, the mineral is classified as gravel.

Block E

This block embraces 14 km^2 of mineral-bearing Younger River Gravels of the River Loddon, which extend southwards from the confluence with the River Thames to the southern margin of the sheet. They lie mainly between 30 m (98.5 ft) and 40 m (131 ft) above OD. They are underlain by Upper Chalk in the north, Reading Beds in the central part of the block and London Clay in the south. About a quarter of the mineral area is covered by Alluvium. The sands and gravels have been exploited extensively to the south of Twyford [7880 7600] and to the north of Winnersh [7815 7050]; at the time of the resurvey there were active workings in both areas.

The assessment of resources is based on 10 IMAU boreholes, 4 Hydrogeological Department records and 85 other boreholes. The mineral has a mean thickness of 2.4 m; it varies from under 1 m locally to 4.9 m in Borehole 268/245. The estimated volume of mineral is 34 million $m^3 \pm 24$ per cent. The overburden averages 1 m in thickness; it ranges from a thin soil cover in some areas to 3 m locally. In the area of mapped Alluvium the overburden has a mean thickness of 1.3 m and in the Alluvium-free areas the mean is 0.8 m.

The fines content (mainly silt) of the mineral varies widely from 6 per cent in Borehole 77 SE 43 to as much as 30 per cent in Borehole 77 NE 27. The sand, which is dominantly medium to coarse in grain size, makes up between 18 and 40 per cent of the deposit. Gravel usually makes up more than half of the mineral, but only 45 per cent was recorded in Borehole 77 SE 45 and 48 per cent in Boreholes 77 SE 44 and 49. The highest gravel content, 74 per cent, was recorded in Borehole 77 SE 47. The mean grading for the block is fines 12 per cent, sand 31 per cent, gravel 57 per cent; overall, the mineral is classified as 'clayey gravel'.

Table 5 Data from assessment boreholes: Block C

Borehole Number	Recorded thickness		Mean grading percentages					
	Mineral	Overburden	Fines	Fine sand	Medium sand	Coarse sand	Fine gravel	Coarse gravel
	(m)	(m)	−1/16 mm	−¼+1/16 mm	−1+¼ mm	−4+1 mm	−16+4 mm	+16 mm
77 NE 22	3.0	0.6	29	10	17	5	13	26
78 SE 17	10.1	0.2	17	1	10	9	24	39
SE 18	5.7	3.6	2	1	12	16	25	44
SE 19	10.8	1.4	10	2	15	11	28	34
SE 20	3.0	2.9	12	2	15	11	28	32
SE 21	8.4	0.9	13	2	25	13	24	23
SE 23	9.1	2.2	18	8	30	7	17	20
SE 24	4.7	2.1	12	3	20	7	24	34
SE 25	8.2	1.7	3	1	14	15	31	36
SE 26	Mineral absent							
SE 27	6.1	0.8	2	1	16	15	36	30
78 NE 9	4.1	0.2	9	1	8	10	25	47
NE 10	3.0	2.3	24	2	18	8	24	34
NE 11	6.5	0.2	5	2	7	12	33	41
NE 12	7.1	0.4	7	2	13	14	31	33
Mean for block			11	5	13	12	25	34

Table 6 Data from assessment boreholes: Block D

Borehole Number	Recorded thickness		Mean grading percentages					
	Mineral	Overburden	Fines	Fine sand	Medium sand	Coarse sand	Fine gravel	Coarse gravel
	(m)	(m)	−1/16 mm	$-\frac{1}{4}+1/16$ mm	$-1+\frac{1}{4}$ mm	−4+1 mm	−16+4 mm	+16 mm
77 SW 8	6.9	3.8	4	2	17	15	25	37
SW 9	6.8	1.3	6	2	15	10	24	43
SW 10	3.3	4.0	5	1	18	10	50	16
SW 11	3.9	1.1	3	2	14	8	47	26
SW 13	7.7	0.7	5	2	13	14	31	35
SW 14	8.3	1.0	6	2	24	17	30	21
77 NE 19	7.3	3.1	5	2	15	19	33	26
NE 20	4.1	3.1	9	0	14	16	27	34
NE 21	4.3	0.4	8	2	21	12	27	30
NE 23	6.6	1.6	6	6	10	13	33	32
NE 24	6.3	0.8	9	4	16	11	5	55
NE 25	2.6	2.2	17	2	9	9	21	42
NE 26	5.9	0.8	4	1	17	11	31	36
Mean for block			6	2	16	14	28	34

Table 7 Data from assessment boreholes: Block E

Borehole Number	Recorded thickness		Mean grading percentages					
	Mineral	Overburden	Fines	Fine sand	Medium sand	Coarse sand	Fine gravel	Coarse gravel
	(m)	(m)	−1/16 mm	$-\frac{1}{4}+1/16$ mm	$-1+\frac{1}{4}$ mm	−4+1 mm	−16+4 mm	+16 mm
77 SE 43	6.6	1.0	6	3	18	9	35	29
SE 44	3.4	0.5	18	4	22	8	28	20
SE 45	3.1	1.0	15	3	26	11	31	14
SE 46	3.9	0.8	9	2	14	14	34	27
SE 47	1.8	1.9	8	1	6	11	42	32
SE 48	3.0	0.4	18	3	19	9	29	22
SE 49	1.3	2.2	14	2	22	14	38	10
SE 50	3.3	0.3	9	2	18	11	38	22
SE 51	2.1	0.5	13	1	13	19	46	8
77 NE 27	1.0	2.0	30	1	10	9	28	22
Mean for block			12	3	17	11	35	22

APPENDIX A: FIELD AND LABORATORY PROCEDURES

Trial and error during initial studies of the complex and variable glacial deposits of East Anglia and Essex showed that an absolute minimum of five sample points evenly distributed across the sand and gravel are needed to provide a worthwhile statistical assessment, but that, where possible, there should be not less than ten. Sample points are any points for which adequate information exists about the nature and thickness of the deposit and may include boreholes other than those drilled during the survey and exposures. In particular, the cooperation of sand and gravel operators ensures that boreholes are not drilled where reliable information is already available; although this may be used in the calculations, it is held confidentially by the Institute and cannot be disclosed.

The mineral shown on each 1:25 000 sheet is divided into resource blocks. The arbitrary size selected, 10 km², is a compromise to meet the aims of the survey by providing sufficient sample points in each block. As far as possible the block boundaries are determined by geological boundaries so that, for example, glacial and river terrace gravels are separated. Otherwise division is by arbitrary lines, which may bear no relationship to the geology. The blocks are drawn provisionally before drilling begins.

A reconnaissance of the ground is carried out to record any exposures and inquiries are made to ascertain what borehole information is available. Borehole sites are then selected to provide an even pattern of sample points at a density of approximately one per square kilometre. However, because broad trends are independently overlain by smaller scale characteristically random variations, it is unnecessary to adhere to a square grid pattern. Thus such factors as ease of access and the need to minimise disturbance to land and the public are taken into account in siting the holes; at the same time it is necessary to guard against the possibility that ease of access (that is, the positions of roads and farms) may reflect particular geological conditions, which may bias the drilling results.

The drilling machine employed should be capable of providing a continuous sample representative of all unconsolidated deposits, so that the in-situ grading can be determined, if necessary, to a depth of 30 m (100 ft) at a diameter of about 200 mm (8 in), beneath different types of overburden. It should be reliable, quiet, mobile and relatively small (so that it can be moved to sites of difficult access). Shell and auger rigs have proved to be almost ideal.

The rigs are modified to enable deposits above the water table to be drilled 'dry', instead of with water added to facilitate the drilling, to minimise the amount of material drawn in from outside the limits of the hole. The samples thus obtained are representative of the in-situ grading, and satisfy one of the most important aims of the survey. Below the water table the rigs are used conventionally, although this may result in the loss of some of the fines fraction and the pumping action of the bailer tends to draw unwanted material into the hole from the sides or the bottom.

A continuous series of bulk samples is taken throughout the sand and gravel. Ideally samples are composed exclusively of the whole of the material encountered in the borehole between stated depths. However, care is taken to discard, as far as possible, material which has caved or has been pumped from the bottom of the hole. A new sample is commenced whenever there is an appreciable lithological change within the sand and gravel, or at every 1 m (3.3 ft) depth. The samples, each weighing between 25 and 45 kg (55 and 100 lb), are despatched in heavy duty polythene bags to a laboratory for grading. The grading procedure is based on British Standard 1377 (1967). Random checks on the accuracy of the grading are made in the Institute's laboratories.

All data, including mean grading analysis figures calculated for the total thickness of the mineral, are entered on standard record sheets, abbreviated copies of which are reproduced in Appendix F.

Detailed records may be consulted at the appropriate offices of the Institute, upon application to the Head, Industrial Minerals Assessment Unit.

APPENDIX B: STATISTICAL PROCEDURE

Statistical Assessment

1. A statistical assessment is made of an area of mineral greater than 2 km², if there is a minimum of five evenly spaced boreholes in the resource block (for smaller areas see paragraph 12 below).

2. The simple methods used in the calculations are consistent with the amount of data provided by the survey. Conventional symmetrical confidence limits are calculated for the 95 per cent probability level, that is, there is a 5 per cent or one in twenty chance of a result falling outside the stated limits.

3. The volume estimate (V) for the mineral in a given block is the product of the two variables, the sampled areas (A) and the mean thickness ($\bar{l}_m$) calculated from the individual thicknesses at the sample points. The standard deviations for these variables are related such that

$$S_V = \sqrt{(S_A{}^2 + S_{\bar{l}_m}{}^2)} \qquad [1]$$

4. The above relationship may be transposed such that

$$S_V = S_{\bar{l}_m} \sqrt{(1 + S_A{}^2/S_{\bar{l}_m}{}^2)} \qquad [2]$$

From this it can be seen that as $S_A{}^2/S_{\bar{l}_m}{}^2$ tends to 0, S_V tends to $S_{\bar{l}_m}$.
If, therefore, the standard deviation for area is small with respect to that for mean thickness, the standard deviation for volume approximates to that for mean thickness.

5. Given that the number of approximately evenly spaced sample points in the sampled area is n with mineral thickness measurements $l_{m_1}, l_{m_2}, \ldots l_{m_n}$, then the best estimate of mean thickness, $\bar{l}_m$, is given by

$$\frac{\sum(\bar{l}m_1 + lm_2 \ldots lm_n)}{n}$$

For groups of closely spaced boreholes a discretionary weighting factor may be applied to avoid bias (see note on weighting below). The standard deviation for mean thickness, $S_{\bar{l}}$, expressed as a proportion of the mean thickness is given by

$$S_{\bar{l}} = (1/\bar{l}_m)\sqrt{[(l_m - \bar{l}_m)^2/(n-1)]}$$

where l_m is any value in the series l_{m_1} to l_{m_n}.

6. The sampled area in each resource block is coloured pink on the map. Wherever possible, calculations relate to the mineral within mapped geological boundaries (which may not necessarily correspond to the limits of deposit). Where the area is not defined by a mapped boundary, that is, where the boundary is inferred, a distinctive symbol is used. Experience suggests that the errors in determining area are small relative to those in thickness. The relationship $S_A/S_{\bar{l}_m} \leqslant ⅓$ is assumed in all cases. It follows from equation [2] that

$$S_{\bar{l}_m} \leqslant S_V \leqslant 1.05\, S_{\bar{l}_m} \qquad [3]$$

7. The limits on the estimate of mean thickness of mineral, $L_{\bar{l}_m}$, may be expressed in absolute units $\pm(t/\sqrt{n}) \times S_{\bar{l}_m}$ or as a percentage $\pm(t/\sqrt{n}) \times S_{\bar{l}_m} \times (100/\bar{l}_m)$ per cent, where t is Student's t at the 95 per cent probability level for $(n-1)$ degrees of freedom, evaluated by reference to statistical tables. (In applying Student's t it is assumed that the measurements are distributed normally).

8. Values of t at the 95 per cent probability level for values of n up to 20 are as follows:

n	t	*n*	t
1	infinity	11	2.228
2	12.706	12	2.201
3	4.303	13	2.179
4	3.182	14	2.160
5	2.776	15	2.145
6	2.571	16	2.131
7	2.447	17	2.120
8	2.365	18	2.110
9	2.306	19	2.101
10	2.262	20	2.093

(from Table 12, Biometrika Tables for Statisticians, Volume 1, Second Edition, Cambridge University Press, 1962). When n is greater than 20, 1.96 is used (the value of t when n is infinity).

9. In calculating confidence limits for volume, L_V, the following inequality corresponding to equation [3] is applied: $L_{\bar{l}_m} \leqslant L_V \leqslant 1.05\, L_{\bar{l}_m}$

10. In summary, for values of n between 5 and 20, L_V is calculated as

$$[(1.05 \times t)/\bar{l}_m] \times [\sqrt{\Sigma(l_m - \bar{l}_m)^2/n(n-1)}] \times 100$$

per cent, and when n is greater than 20, as

$$[(1.05 \times 1.96)/\bar{l}_m] \times [\sqrt{\Sigma(l_m - \bar{l}_m)^2/n(n-1)}] \times 100$$

per cent.

11. The application of this procedure to a fictitious area is illustrated in Figs. 5 and 6

12. If the sampled area of mineral in a resource block is between 0.25 km^2 and 2 km^2 an assessment is inferred, based on geological and topographical information usually supported by the data from one or two boreholes. The volume of mineral is calculated as the product of the area, measured from field data, and the estimated thickness. Confidence limits are not calculated.

13. In some cases a resource block may include an area left uncoloured on the map, within which mineral (as defined) is interpreted to be generally absent. If there is reason to believe that some mineral may be present, an inferred assessment may be made.

14. No assessment is attempted for an isolated area of mineral less than 0.25 km^2.

15. *Note on Weighting* The thickness of a deposit at any point may be governed solely by the position of the point in relation to a broad trend. However, most sand and gravel deposits also exhibit a random pattern of local, and sometimes considerable, variation in thickness. Thus the distribution of sample points need be only approximately regular and in estimating the mean thickness only simple weighting is necessary. In practice, equal weighting can often be applied to thicknesses at all sample points. If, however, there is a distinctly unequal distribution of points, bias is avoided by dividing the sampled area into broad zones, to each of which a value roughly proportional to its area is assigned. This value is then shared between the data points within the zone as the weighting factor.

APPENDIX C: CLASSIFICATION AND DESCRIPTION OF SAND AND GRAVEL

For the purposes of assessing resources of sand and gravel a classification should take account of economically important characteristics of the deposit, in particular the absolute content of fines and the ratio of sand to gravel.

The terminology commonly used by geologists when describing sedimentary rocks (Wentworth, 1922) is not entirely satisfactory for this purpose. For example, Wentworth proposed that a deposit should be described as a 'gravelly sand' when it contains more sand than gravel and there is at least 10 per cent of gravel, provided that there is less than 10 per cent of material finer than sand (less than $\frac{1}{16}$ mm) and coarser than pebbles (more than 64 mm in diameter). Because deposits containing more than 10 per cent fines are not embraced by this system a modified binary classification based on Willman (1942) has been adopted.

When the fines content exceeds 40 per cent the material is not considered to be potentially workable and falls outside the definition of mineral. Deposits which contain 40 per cent fines or less are classified primarily on the ratio of sand to gravel but qualified in the light of the fines content, as follows: less than 10 per cent fines – no qualification; 10 per cent or more but less than 20 per cent fines – 'clayey'; 20 to 40 per cent fines – 'very clayey'.

The term 'clay' (as written, with single quote marks) is used to describe all material passing $\frac{1}{16}$ mm. Thus it has no mineralogical significance and includes particles falling within the size range of silt. The normal meaning applies to the term clay where it does not appear in single quotation marks.

The ratio of sand to gravel defines the boundaries between sand, pebbly sand, sandy gravel and gravel (at 19:1, 3:1 and 1:1).

Thus it is possible to classify the mineral into one of twelve descriptive categories (see Fig. 7). The procedure is as follows:

1. Classify according to ratio of sand to gravel.
2. Describe fines.

For example, a deposit grading 11 per cent gravel, 70 per cent sand and 19 per cent fines is classified as 'clayey' pebbly sand. This short description is included in the borehole log (see Note 11, p. 20)

Many differing proposals exist for the classification of the grain size of sediments (Atterberg, 1905; Udden, 1914; Wentworth, 1922; Wentworth, 1935; Allen, 1936; Twenhofel, 1937; Lane and others, 1947). As Archer (1970a, b) has emphasised, there is a pressing need for a simple metric scale acceptable to both scientific and engineering interests, for which the class limit sizes correspond closely with certain marked changes in the natural properties of mineral particles. For example, there is an important change in the degree of cohesion between particles at about the $\frac{1}{16}$-mm size, which approximates to the generally accepted boundary between silt and sand. These and other requirements are met by a system based on Udden's geometric scale and a simplified form of Wentworth's terminology (Table 8), which is used in this Report.

The fairly wide intervals in the scale are consistent with the general level of accuracy of the qualitative assessments of the resource blocks. Three sizes of sand are recognised, fine ($-\frac{1}{4}$ $+\frac{1}{16}$ mm), medium (-1 $+\frac{1}{4}$ mm) and coarse (-4 $+1$ mm). The boundary at 16 mm distinguishes a range of finer gravel (-16 $+4$ mm), often characterised by abundance of worn tough pebbles of vein quartz, from larger pebbles often of notably different materials. The boundary at 64 mm distinguishes pebbles from cobbles. The term 'gravel' is used loosely to denote both pebble-sized and cobble-sized material.

The size distribution of borehole samples is determined by sieve analysis, which is presented by the laboratory as logarithmic cumulative curves (see, for example, British Standard 1377: 1967). In this report the grading is tabulated on the borehole record sheets (Appendix F), the intercepts corresponding with the simple geometric scale $\frac{1}{16}$ mm, $\frac{1}{4}$ mm, 1 mm, 4 mm, 16 mm and so on as required. Original sample grading curves are available for reference at the appropriate office of the Institute.

Each bulk sample is described, subjectively, by a geologist at the borehole site. Being based on visual examination, the description of the grading is inexact, the accuracy depending on the experience of the observer. The descriptions recorded are modified, as necessary, when the laboratory results become available.

The relative proportions of the rock types present in the gravel fraction are indicated by the use of the words 'and' or 'with'. For example, 'flint and quartz' indicates very approximate equal proportions with neither constituent accounting for less than about 25 per cent of the whole; 'flint with quartz' indicates that flint is dominant and quartz, the principal accessory rock types, comprises 5 to 25 per cent of the whole. Where the accessory material accounts for less than 5 per cent of the whole, but is still readily apparent, the phrase 'with some' has been used. Rare constituents are referred to as 'trace'.

The terms used in the field to describe the degree of rounding of particles, which is concerned with the sharpness of the edges and corners of a clastic fragment and not the shape (after Pettijohn, 1957), are as follows.

Angular: showing little or no evidence of wear; sharp edges and corners.

Subangular: showing definite effects of wear. Fragments still have their original form but edges and corners begin to be rounded off.

Subrounded: showing considerable wear. The edges and corners are rounded off to smooth curves. Original grain shape is still distinct.

Rounded: original faces almost completely destroyed, but some comparatively flat surfaces may still remain. All original edges and corners have been smoothed off to rather broad curves. Original shape is still apparent.

Well-rounded: no original faces, edges or corners left. The entire surface consists of broad curves; flat areas are absent. The original shape is suggested by the present form of the grain.

Table 8 Classification of gravel, sand and fines

Size limits	Grain size description	Qualification	Primary classification
	Cobble		
64 mm			
	Pebble	Coarse	Gravel
16 mm			
		Fine	
4 mm			
		Coarse	
1 mm			
	Sand	Medium	Sand
$\frac{1}{4}$ mm			
		Fine	
$\frac{1}{16}$ mm			
	Fines (silt and clay)		Fines

Block Calculation 1:25 000 Block } Fictitious

Area
- Block: 11.08 km^2
- Mineral: 8.32 km^2

Volume
- Overburden: 21 million m^3
- Mineral: 54 million m^3

Mean Thickness
- Overburden: 2.5 m
- Mineral: 6.5 m

Confidence limits of the estimate of mineral volume at the 95 per cent probability level: ± 20 per cent

That is, the volume of mineral (with 95 per cent probability): 54 ± 11 million m^3

Thickness estimate: measurements in metres

l_o = overburden thickness l_m = mineral thickness

Sample point	Weighting w	Overburden l_o	wl_o	Mineral l_m	wl_m	Remarks
SE 14	1	1.5	1.5	9.4	9.4	IMAU boreholes
SE 18	1	3.3	3.3	5.8	5.8	
SE 20	1	nil	-	6.9	6.9	
SE 22	1	0.7	0.7	6.4	6.4	
SE 23	1	6.2	6.2	4.1	4.1	
SE 24	1	4.3	4.3	6.4	6.4	
SE 17	½	1.2	1.6	9.8	7.2	Hydrogeological Dept record
123/45	½	2.0		4.6		
1	¼	2.7	2.6	7.3	5.8	Close group of four boreholes (commercial)
2	¼	4.5		3.2		
3	¼	0.4		6.8		
4	¼	2.8		5.9		
Totals	$\Sigma w = 8$	$\Sigma wl_o = 20.2$		$\Sigma wl_m = 52.0$		
Means		$\bar{l}_o = 2.5$		$\bar{l}_m = 6.5$		

Calculation of confidence limits

l_m	$(l_m - \bar{l}_m)$	$(l_m - \bar{l}_m)^2$
9.4	2.9	8.41
5.8	0.7	0.49
6.9	0.4	0.16
6.4	0.1	0.01
4.1	2.4	5.76
6.4	0.1	0.01
7.2	0.7	0.49
5.8	0.7	0.49

$\Sigma(l_m - \bar{l}_m)^2 = 15.82$

n = 8

t = 2.365

L_V is calculated as

$$1.05 \times \frac{t}{\bar{l}_m} \sqrt{\frac{\Sigma(l_m - \bar{l}_m)^2}{n(n-1)}} \times 100$$

$$= 1.05 \times \frac{2.365}{6.5} \sqrt{\frac{15.82}{8 \times 7}} \times 100$$

$$= 20.3$$

$$\simeq 20 \text{ per cent}$$

Figure 5. Example of resource block assessment, calculations and results

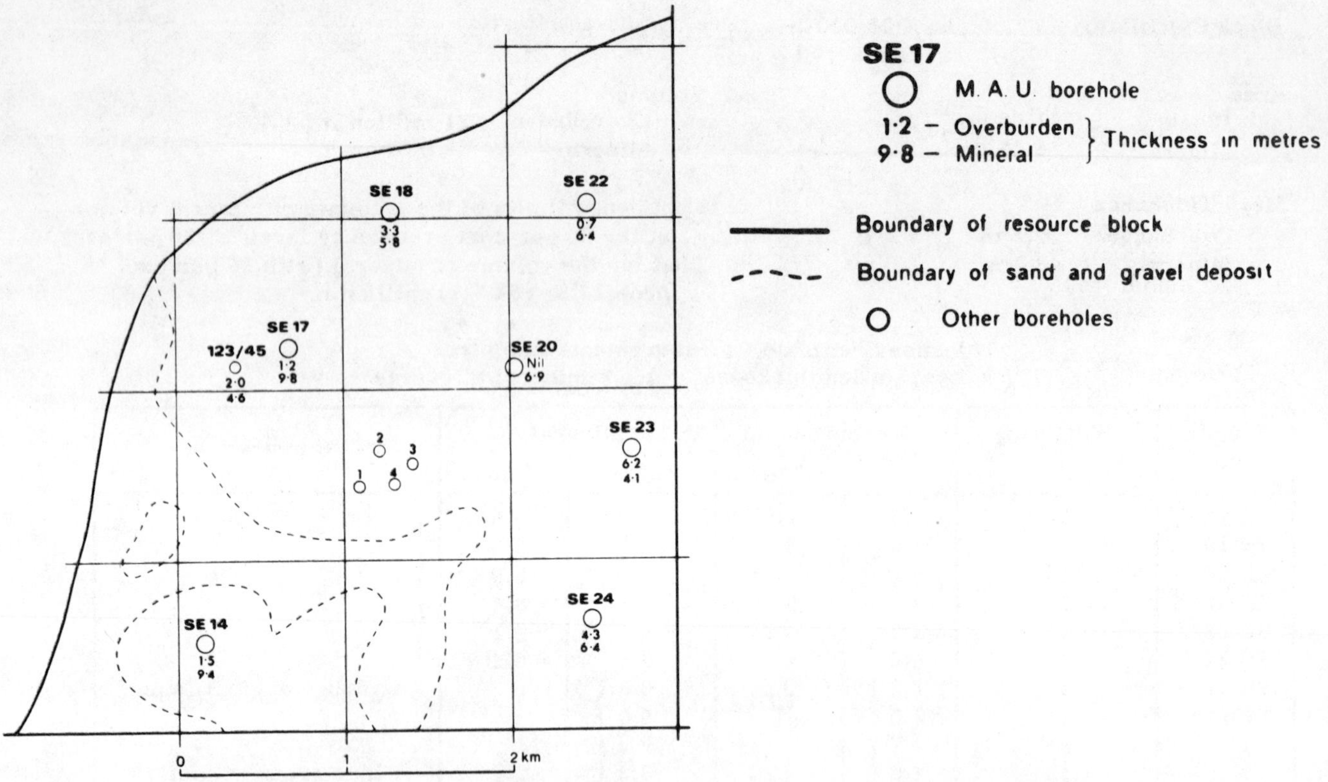

Fig. 6 Example of resource block assessment: map of a fictitious block

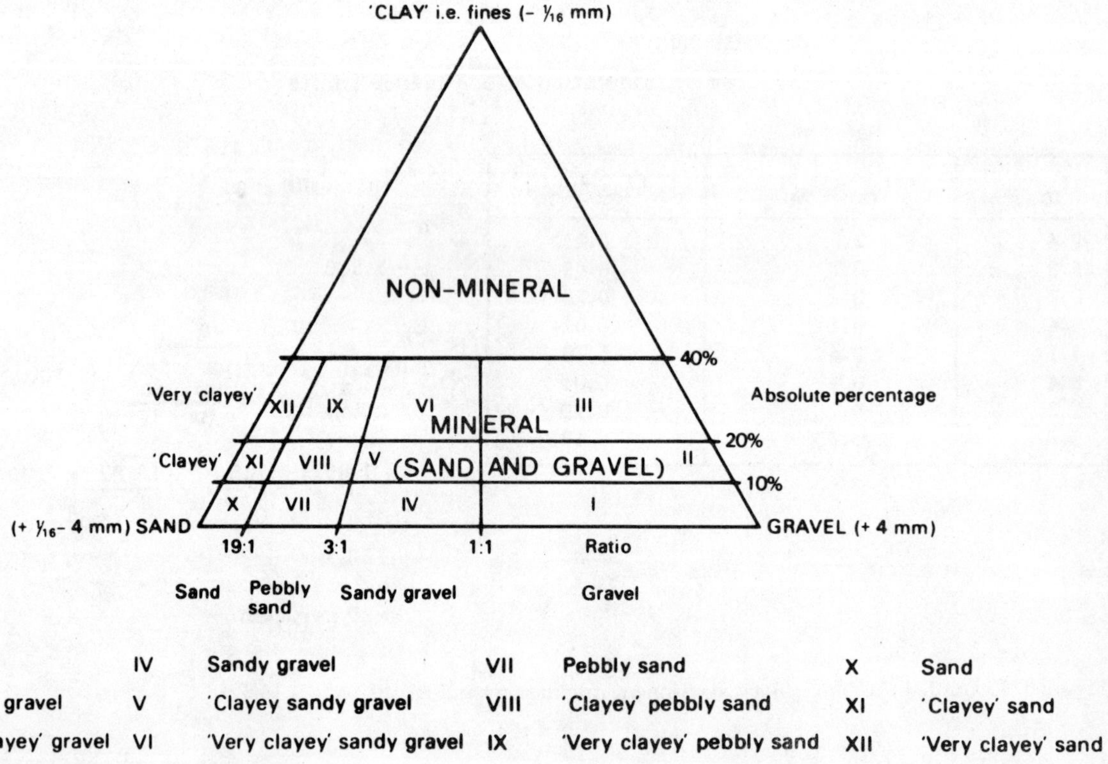

Fig. 7 Diagram to show the descriptive categories used in the classification of sand and gravel

APPENDIX D: EXPLANATION OF THE BOREHOLE RECORDS

ANNOTATED EXAMPLE

SU 77 NE 23[1] 7752 7796[2] Borough Marsh, near Wargrove[3] Block D

Surface level +35.0 m (+115 ft)[4]
Water struck at +32.4 m[5]
Shell and auger (modified) 6 in (152 mm) diameter[6]
December 1972

Overburden 1.6 m (5.5 ft)[7]
Mineral 6.6 m (21.5 ft)
Bedrock 0.5 m+ (1.5 ft+)

LOG

Geological classification	Lithology	Thickness m	Thickness (ft)	Depth m	Depth (ft)[8]
	Soil	0.3	(1.0)	0.3	(1.0)
Alluvium[10]	[11]Clay, brown, sandy, with scattered flint pebbles	1.3	(4.5)	1.6	(5.5)
Younger River Gravels (undifferentiated)	Gravel Gravel: fine to coarse with occasional cobbles, subangular to subrounded flint, with limestone and some quartz, quartzite and a trace of chalk and ironstone Sand: mainly coarse and medium	6.6	(21.5)	8.2	(27.0)
Upper Chalk	Chalk with flint nodules	0.5+[9]	(1.5+)	8.7	(28.5)

GRADING

Mean for deposit

	%	mm	%
[15]Gravel	55	+16	32
		−16+4	33
Sand	29	−4+1	13
		−1+¼	10
		−¼+1/16	6
Fines	6	−1/16	6

Bulk samples

Depth below surface (m)	Percentages Fines	Sand	Gravel
1.6 − 2.6[12]	26	36	38[13]
[14]*2.6 − 3.6	14	36	50
*3.6 − 4.6	2	10	88
*4.6 − 5.6	1	18	81
*5.6 − 6.6	11	32	57
*6.6 − 7.6	6	30	64
*7.6 − 8.2	2	16	82

COMPOSITION[16]

Depth below surface (m)	Percentage (in 4.75 to 9.5 mm fractions) Flint	Quartz	Ironstone	Chalk	Limestone
1.6 − 2.6	76	5	3	3	13
2.6 − 3.6	70	6	2	3	19
3.6 − 4.6	70	12	4	2	11
4.6 − 5.6	75	7	1	2	14
5.6 − 6.6	77	5	3	2	13
6.6 − 7.6	79	6	3	4	8
7.6 − 8.2	83	6	2	3	6
Mean	75	7	3	3	12

The numbered paragraphs below correspond with the annotations given on the specimen record above.

1 Borehole Registration Number
Each Industrial Minerals Assessment Unit (IMAU) borehole is identified by a Registration Number. This consists of two statements.

1) The number of the 1:25 000 sheet on which the borehole lies, for example SU 78.
2) The quarter of the 1:25 000 sheet on which the borehole lies and its number in a series for that quarter, for example SE 20.

Thus the full Registration Number is SU 78 SE 20. Usually this is abbreviated to 78 SE 20 in the text.

2 The National Grid Reference
All National Grid References in this publication lie within the 100-km square SU unless otherwise stated. Grid references are given to eight figures, accurate to within 10 m for borehole locations. (In the text, six-figure grid references are used for more approximate locations, for example, for farms).

3 Location
The position of the borehole is generally referred to the nearest named locality on the 1:25 000 base map and the resource block in which it lies is stated.

4 Surface level
The surface level at the borehole site is given in metres and feet above Ordnance Datum. Measurements were made in either metres or feet; approximate conversions to feet or metres are given in brackets.

5 Groundwater conditions
If groundwater was present the level at which it was encountered is normally given (in metres above OD).

6 Type of drill and Date of drilling
Modified shell and auger rigs were used in this survey. The type of machine, the external diameter of the casing used, and the month and year of completion of the borehole are stated.

7 Overburden, Mineral, Waste and Bedrock
Mineral is sand and gravel which, as part of a deposit, falls within the arbitrary definition of potentially workable material (see p. 1). Bedrock is the 'formation', 'country rock' or 'rock head' below which potentially workable sand and gravel will not be found. Waste is any material other than bedrock or mineral. Where waste occurs between the surface and mineral it is classified as overburden.

8 Thickness and Depth
All measurements were made in metres; imperial conversions appear in brackets. Imperial conversions of measurements of the thicknesses of beds and the depths from the surface of their bases have been rounded off to the nearest 0.5 ft, because a more detailed quotation would imply a higher order of accuracy than could be justified by the original figures. Where figures have been rounded in this way there may be a discrepancy between the sum of the thicknesses and the recorded depths.

9 The plus sign (+) indicates that the base of the deposit was not reached during drilling.

10 Geological classification
The geological classification is given whenever possible.

11 Lithological description
When sand and gravel is recorded a general description based on the mean grading characteristics (for details see Appendix C) is followed by more detailed particulars. The description of other rocks is based on visual examination, in the field.

12 Sampling
A continuous series of bulk samples is taken throughout the thickness of sand and gravel. A new sample is commenced whenever there is an appreciable lithological change within the sand and gravel or at every 1 m of depth.

13 Grading results
The limits are as follows: gravel, +4 mm; sand, −4+1/16 mm; fines, −1/16 mm.

14 Bailed samples
Samples obtained by the bailing technique (that is, from deposits below the water table) are indicated by an asterisk.

15 Mean grading
The grading of the full thickness of the mineral horizon identified in the log is the mean of the individual sample gradings weighted by the thicknesses represented, if these vary. The classification used is shown in Table 8.

Fully representative sampling of sand and gravel is difficult to achieve particularly where groundwater levels are high. Comparison between boreholes and adjacent exposures suggests that in borehole samples the proportion of sand may be higher and the proportion of fines and coarse gravel (+16 mm) may be lower.

16 Composition
Where deleterious material is present in substantial amounts, details of the compositions (percentages by weight of each constituent) of the 4.75 to 9.5-mm fractions of bulk samples are given.

APPENDIX E: LIST OF BOREHOLES USED IN ASSESSMENT OF RESOURCES

INDUSTRIAL MINERALS ASSESSMENT UNIT BOREHOLES

Borehole number by sheet quadrant		Grid references (all fall in 100 km square SU)	Page No.
77 NW	50	7042 7907	22
	51	7017 7756	23
	52	7113 7940	24
	53	7098 7792	25
	54	7205 7971	26
	55	7171 7883	27
	56	7308 7973	28
	57	7452 7954	29
	58	7424 7875	30
	59	7410 7762	30
77 NE	17	7590 7794	31
	18	7661 7849	32
	19	7664 7775	33
	20	7609 7672	34
	21	7678 7521	35
	22	7734 7963	36
	23	7752 7796	37
	24	7791 7647	38
	25	7712 7600	39
	26	7807 7906	40
	27	7849 7728	41
77 SW	8	7044 7475	42
	9	7037 7431	43
	10	7080 7104	44
	11	7029 7035	45
	12	7236 7402	46
	13	7353 7487	47
	14	7395 7428	48
77 SE	41	7582 7465	49
	42	7673 7424	50
	43	7665 7155	51
	44	7609 7054	52
	45	7664 7029	53
	46	7707 7127	54
	47	7853 7432	55
	48	7882 7244	56
	49	7950 7430	57
	50	7915 7361	58
	51	7944 7181	59
78 NW	8	7054 8645	60
	9	7054 8534	60
	10	7160 8871	61
	11	7163 8596	61
	12	7248 8552	62
	13	7493 8583	62
78 NE	7	7597 8842	63
	8	7572 8711	63
	9	7768 8952	64
	10	7740 8551	65
	11	7818 8740	66
	12	7858 8509	67
	13	7934 8894	68
	14	7977 8761	68
	15	7966 8600	69
78 SW	4	7030 8419	69
	5	7042 8156	70
	6	7033 8058	71
	7	7164 8478	72
	8	7115 8308	73
	9	7127 8234	73
	10	7190 8045	74
	11	7279 8471	75
	12	7249 8395	75
	13	7219 8278	76
	14	7265 8200	77
	15	7325 8326	78
	16	7339 8205	78
	17	7450 8307	79
	18	7451 8150	79
	19	7472 8039	80
78 SE	14	7550 8432	81
	15	7537 8161	82
	16	7565 8017	83
	17	7633 8480	84
	18	7684 8430	85
	19	7623 8356	86
	20	7679 8261	87
	21	7749 8456	88
	22	7778 8317	89
	23	7728 8148	90
	24	7721 8068	91
	25	7812 8486	92
	26	7821 8068	93
	27	7972 8399	94
	28	7971 8245	95
	29	7936 8162	95

OTHER BOREHOLES

1 Hydrogeological Department: 268/109, 242, 244 c, 245, 295, 419, 427, 432 and 476; 254/1 a, 62, 150, 366 b, 373 and 435.

2 Other IGS–registered boreholes: 77 SE 3, 6, 7, 8, 9, 10, 11, 12, 13, 15, 16, 17, 18, 19, 20, 21, 22, 23, 25, 26, 27, 28, 29, 30, 31, 32, 34, 35, 36, 37, 38, 39 and 40.

3 Sand and Gravel Industry and Local Authorities: many records are held in confidence.

APPENDIX F: INDUSTRIAL MINERALS ASSESSMENT UNIT BOREHOLE RECORDS

SU 77 NW 50 7042 7907 Vines Farm, Kidmore End

Surface level (+ 88.7 m) + 290 ft
Water not struck
Shell and auger (modified) 6 in (152 mm) diameter
November 1972

Waste 4.4 m (14.5 ft)
Bedrock 0.5 m+ (1.5 ft+)

Geological Classification	LOG Lithology	Thickness m	 ft	Depth m	 ft
	Soil	0.4	(1.5)	0.4	(1.5)
Older River Gravels (undifferentiated)	Clay, reddish brown to dark brown, silty and sandy, with some flint cobbles and well rounded quartz and quartzite	4.0	(13.0)	4.4	(14.5)
Upper Chalk	Chalk	0.5+	(1.5+)	4.9	(16.0)

SU 77 NW 51 | 7017 7756 | Tokers Green, Kidmore End | Block B

Surface level (+ 79.9 m) + 262 ft
Water not struck
Shell and auger (modified) 6 in (152 mm) diameter
November 1972

Overburden 0.2 m (0.5 ft)
Mineral 3.0 m (10.0 ft)
Waste 0.8 m (2.5 ft)
Bedrock 0.5 m + (1.5 ft+)

LOG

Geological Classification	Lithology	Thickness m	ft	Depth m	ft
	Soil	0.2	(0.5)	0.2	(0.5)
Older River Gravels (undifferentiated)	'Very clayey' sandy gravel Gravel: fine to coarse with occasional cobbles, subangular to well rounded flint, with some well rounded quartz and some subangular to well rounded quartzite Sand: mainly medium with a little coarse and fine, clayey throughout, yellow with ferruginous staining	3.0	(10.0)	3.2	(10.5)
	Clay, very sandy with some coarse and fine gravel	0.8	(2.5)	4.0	(13.0)
Upper Chalk	Chalk	0.5+	(1.5+)	4.5	(15.0)

GRADING

Mean for deposit

	%	mm	%
Gravel	31	+16	15
		−16+4	16
Sand	42	−4+1	7
		−1+¼	30
		−¼+1/16	5
Fines	27	−1/16	27

Bulk samples

Depth below surface (m)	Percentages Fines	Sand	Gravel
0.2 − 1.2	38	24	38
1.2 − 2.2	23	45	32
2.2 − 3.2	19	60	21

SU 77 NW 52 | 7113 7940 | Kennylands, Sonning Common | Block B

Surface level (+ 81.7 m) + 268 ft
Water not struck
Shell and auger (modified) 6 in (152 mm) diameter
November 1972

Overburden 0.3 m (1.0 ft)
Mineral 5.9 m (19.5 ft)
Bedrock 0.5 m+ (1.5 ft+)

LOG

Geological Classification	Lithology	Thickness m	ft	Depth m	ft
	Soil	0.3	(1.0)	0.3	(1.0)
Older River Gravels (undifferentiated)	'Very clayey' gravel Gravel: mainly coarse with some fine and occasional cobbles, angular to well rounded brown and black flint, with some well rounded quartz and quartzite and sandstone Sand: medium with some coarse, clayey throughout, brown	5.9	(19.5)	6.2	(20.5)
Upper Chalk	Chalk	0.5+	(1.5+)	6.7	(22.0)

GRADING

Mean for deposit

	%	mm	%
Gravel	52	+16	37
		−16+4	15
Sand	28	−4+1	7
		$-1+\frac{1}{4}$	20
		$-\frac{1}{4}+1/16$	1
Fines	20	−1/16	20

Bulk samples

Depth below surface (m)	Percentages Fines	Sand	Gravel
0.3 − 1.3	25	32	43
1.3 − 2.3	38	32	30
2.3 − 3.3	26	43	31
3.3 − 4.3	11	24	65
4.3 − 5.3	6	16	78
5.3 − 6.2	11	25	64

SU 77 NW 53 7098 7792 North of Tanner's Farm, Kidmore End Block B

Surface level (+ 79.3 m) + 260 ft
Water not struck
Shell and auger (modified) 6 in (152 mm) diameter
November 1972

Overburden 0.2 m (0.5 ft)
Mineral 4.5 m (15.0 ft)
Bedrock 0.5 m + (1.5 ft+)

LOG

Geological Classification	Lithology	Thickness m	ft	Depth m	ft
	Soil	0.2	(0.5)	0.2	(0.5)
Older River Gravels (undifferentiated)	'Clayey' gravel Gravel: fine to coarse with occasional cobbles, angular to well rounded flint with some well rounded quartz and quartzite Sand: medium with coarse and a little fine, clayey throughout, brown to light brown	4.5	(15.0)	4.7	(15.5)
Upper Chalk	Chalk	0.5+	(1.5+)	5.2	(17.0)

GRADING

Mean for deposit

	%	mm	%
Gravel	50	+16	20
		−16+4	30
Sand	37	−4+1	9
		$-1+\frac{1}{4}$	25
		$-\frac{1}{4}+1/16$	3
Fines	13	−1/16	13

Bulk samples

Depth below surface (m)	Percentages Fines	Sand	Gravel
0.2 − 1.2	15	37	48
1.2 − 2.2	16	34	50
2.2 − 3.2	11	49	40
3.2 − 4.2	12	30	58
4.2 − 4.7	11	29	60

SU 77 NW 54	7205 7971	Frieze Farm, Shiplake	Block B

Surface level (+ 80.5 m) + 264 ft
Water not struck
Shell and auger (modified) 6 in (152 mm) diameter
November 1972

Overburden 0.1 m (0.5 ft)
Mineral 6.2 m (20.5 ft)
Bedrock 0.5 m + (1.5 ft +)

LOG

Geological Classification	Lithology	Thickness m	ft	Depth m	ft
	Soil	0.1	(0.5)	0.1	(0.5)
Older River Gravels (undifferentiated)	'Clayey' gravel Gravel: coarse with some fine, subrounded to well rounded flint with some well rounded quartz and quartzite Sand: mainly medium with some coarse and a little fine, clayey particularly near the top of the mineral, brown	6.2	(20.5)	6.3	(20.5)
Upper Chalk	Chalk with flints	0.5 +	(1.5 +)	6.8	(22.5)

GRADING

Mean for deposit

	%	mm	%
Gravel	48	+16	18
		−16+4	30
Sand	35	−4+1	11
		$-1+\frac{1}{4}$	20
		$-\frac{1}{4}+1/16$	4
Fines	17	−1/16	17

Bulk samples

Depth below surface (m)	Percentages Fines	Sand	Gravel
0.1 − 0.8	41	28	31
0.8 − 1.8	19	35	46
1.8 − 2.8	15	33	52
2.8 − 3.8	17	36	47
3.8 − 4.8	7	32	61
4.8 − 6.3	13	40	47

SU 77 NW 55 7171 7883 East of Cucumber Plantation, Sonning Common Block B

Surface level (+ 80.2 m) + 263 ft
Water not struck
Shell and auger (modified) 6 in (152 mm) diameter
November 1972

Overburden 0.2 m (0.5 ft)
Mineral 7.1 m (23.5 ft)
Bedrock 0.5 m+ (1.5 ft+)

LOG

Geological Classification	Lithology	Thickness m	Thickness ft	Depth m	Depth ft
	Soil	0.2	(0.5)	0.2	(0.5)
Older River Gravels (undifferentiated)	'Clayey' gravel Gravel: fine to coarse with occasional cobbles, subangular to well rounded flint with some well rounded quartz and quartzite Sand: medium with some coarse, very clayey to 1.3 m but becoming almost clayfree between 2.3 m and 5.4 m, light brown	7.1	(23.5)	7.3	(24.0)
Upper Chalk	Chalk	0.5+	(1.5+)	7.8	(25.5)

GRADING

Gravel	50	+16	27
		−16+4	23
Sand	32	−4+1	10
		$-1+\frac{1}{4}$	20
		$-\frac{1}{4}+1/16$	2
Fines	18	−1/16	18

Bulk samples

Depth below surface (m)	Percentages Fines	Sand	Gravel
0.2 − 1.3	40	19	41
1.3 − 2.3	24	24	52
2.3 − 3.3	9	25	66
3.3 − 4.3	7	27	66
4.3 − 5.4	8	23	69
5.4 − 6.4	19	52	29
6.4 − 7.3	17	44	39

SU 77 NW 56 7308 7973 Crowsley Park, Shiplake Block B

Surface level (+ 78.3 m) + 257 ft
Water not struck
Shell and auger (modified) 6 in (152 mm) diameter
December 1972

Overburden 0.1 m (0.5 ft)
Mineral 4.7 m (15.5 ft)
Bedrock 0.5 m + (1.5 ft+)

LOG

Geological Classification	Lithology	Thickness m	ft	Depth m	ft
	Soil	0.1	(0.5)	0.1	(0.5)
Older River Gravels (undifferentiated)	'Clayey' gravel Gravel: fine with some coarse, subangular to subrounded flint with well rounded quartz and some well rounded quartzite Sand: medium with some coarse, fairly clayey throughout, brown	4.7	(15.5)	4.8	(15.5)
Upper Chalk	Chalk	0.5+	(1.5+)	5.3	(17.5)

GRADING

Mean for deposit

	%	mm	%
Gravel	46	+16	19
		−16+4	27
Sand	36	−4+1	11
		$-1+\frac{1}{4}$	23
		$-\frac{1}{4}+1/16$	2
Fines	18	−1/16	18

Bulk samples

Depth below surface (m)	Percentages Fines	Sand	Gravel
0.1 – 1.1	25	28	47
1.1 – 2.1	21	39	40
2.1 – 3.1	13	37	50
3.1 – 4.1	14	38	48
4.1 – 4.8	15	38	47

SU 77 NW 57 7452 7954 Near Mays Green, Shiplake Block B

Surface level (+ 89.0 m) + 292 ft
Water not struck
Shell and auger (modified) 6 in (152 mm) diameter
November 1972

Overburden 1.2 m (4.0 ft)
Mineral 2.7 m (9.0 ft)
Waste 6.7 m (22.0 ft)
Bedrock 2.4 m + (8.0 ft +)

LOG

Geological Classification	Lithology	Thickness m	Thickness ft	Depth m	Depth ft
	Soil	0.1	(0.5)	0.1	(0.5)
Older River Gravels (undifferentiated)	Clay, very gravelly, gravel more coarse than fine	1.1	(3.5)	1.2	(4.0)
	'Clayey' gravel	2.7	(9.0)	3.9	(13.0)
	Gravel: fine to coarse, subangular to well rounded flint with some well rounded quartz and quartzite				
	Sand: mainly coarse and medium, clayey in lowest 1.7 m, brown to dark brown				
	Clay, sandy and gravelly, gravel more fine than coarse	6.7	(22.0)	10.6	(35.0)
Reading Beds	Clay, sandy throughout, mottled brown, light brown and grey	2.4+	(8.0+)	13.0	(42.5)

GRADING

Mean for deposit

	%	mm	%
Gravel	63	+16	30
		−16+4	33
Sand	25	−4+1	13
		$-1+\frac{1}{4}$	9
		$-\frac{1}{4}+1/16$	3
Fines	12	−1/16	12

Bulk samples

Depth below surface (m)	Percentages Fines	Sand	Gravel
1.2 − 2.2	5	22	73
2.2 − 3.2	21	22	57
3.2 − 3.9	11	30	59

SU 77 NW 58 7424 7875 Near Coppice Farm, Shiplake Block B

Surface level (+ 93.6 m) + 307 ft
Water struck at + 88.8 m
Shell and auger (modified) 6 in (152 mm) diameter
November 1972

Overburden 2.3 m (7.5 ft)
Mineral 5.2 m (17.0 ft)
Bedrock 0.5 m+ (1.5 ft+)

LOG

Geological Classification	Lithology	Thickness m	ft	Depth m	
	Soil	0.4	(1.5)	0.4	(1.5)
Older River Gravels (undifferentiated)	Clay with sand and trace of gravel	1.9	(6.0)	2.3	(7.5)
	'Very clayey' pebbly sand Gravel: fine with some coarse, subangular to subrounded flint with some well rounded quartz and quartzite Sand: mainly medium, clayey, orange brown to grey	5.2	(17.0)	7.5	(24.5)
Reading Beds	Clay, mottled grey, brown and red	0.5+	(1.5+)	8.0	(26.0)

GRADING

Mean for deposit

	%	mm	%
Gravel	12	+16	3
		−16+4	9
Sand	64	−4+1	9
		$-1+\frac{1}{4}$	47
		$-\frac{1}{4}+1/16$	8
Fines	24	−1/16	24

Bulk samples

Depth below surface (m)	Percentages Fines	Sand	Gravel
2.3 − 3.3	23	62	15
3.3 − 4.3	22	65	13
* 4.3 − 5.3	22	70	8
* 5.3 − 6.3	25	70	5
* 6.3 − 7.5	29	54	17

SU 77 NW 59 7410 7762 Near Bint's Farm, Eye and Dunsden

Surface level (+ 95.7 m) + 314 ft
Water not struck
Shell and auger (modified) 6 in (152 mm) diameter
November 1972

Waste 0.3 m (1.0 ft)
Bedrock 3.7 m+ (12.0 ft+)

LOG

Geological Classification	Lithology	Thickness m	ft	Depth m	ft
	Soil	0.3	(1.0)	0.3	(1.0)
Reading Beds	Clay, mottled brown and grey, sandy, trace of flint pebbles at top	3.7+	(12.0+)	4.0	(13.0)

SU 77 NE 17	7590 7794	Near Shiplakecourt Farm, Shiplake	Block B

Surface level +62.2 m (+ 204 ft)
Water not struck
Shell and auger (modified) 6 in (152 mm) diameter
December 1972

Overburden 0.4 (1.5 ft)
Mineral 4.7 m (15.5 ft)
Bedrock 0.5 m+ (1.5 ft+)

LOG

Geological Classification	Lithology	Thickness m	ft	Depth m	ft
	Soil	0.4	(1.5)	0.4	(1.5)
Older River Gravels (undifferentiated)	'Very clayey' gravel with a band of clay from 2.1 to 2.8 m Gravel: fine to coarse with some cobbles, subrounded to well rounded flint with subrounded to well rounded quartz and quartzite Sand: Coarse to fine, clayey throughout, brown	4.7	(15.5)	5.1	(17.0)
Upper Chalk	Chalk with flint nodules	0.5+	(1.5+)	5.6	(18.5)

GRADING

Mean for deposit

	%	mm	%
Gravel	43	+16	34
		−16+4	9
Sand	22	−4+1	7
		$-1+\frac{1}{4}$	11
		$-\frac{1}{4}+1/16$	4
Fines	35	−1/16	35

Bulk samples

Depth below surface (m)	Percentages Fines	Sand	Gravel
0.4 − 1.3	35	30	35
1.3 − 2.1	58	15	27
2.8 − 3.8	30	22	48
3.8 − 5.1	25	30	45

SU 77 NE 18 7661 7849 Shiplake Cross, Shiplake Block B

Surface level +60.0 m (+197 ft)
Water not struck
Shell and auger (modified) 6 in (152 mm) diameter
December 1972

Overburden 1.1 m (3.5 ft)
Mineral 4.7 m (15.5 ft)
Bedrock 0.5 m+ (1.5 ft+)

LOG

Geological Classification	Lithology	Thickness m	Thickness ft	Depth m	Depth ft
	Soil	0.2	(0.5)	0.2	(0.5)
Older River Gravels (undifferentiated)	Clay, with some flint and quartzite gravel	0.9	(3.0)	1.1	(3.5)
	'Clayey' gravel Gravel: fine to coarse with occasional cobbles, mainly sub-angular to well rounded flint with some well rounded quartz and quartzite and occasional sandstone Sand: mainly medium with some coarse, clayey throughout, light brown to yellowish brown	4.7	(15.5)	5.8	(19.0)
Upper Chalk	Chalk	0.5+	(1.5+)	6.3	(20.5)

GRADING

Mean for deposit

	%	mm	%
Gravel	53	+16	25
		−16+4	28
Sand	32	−4+1	11
		$-1+\frac{1}{4}$	21
		$-\frac{1}{4}+1/16$	0
Fines	15	−1/16	15

Bulk samples

Depth below surface (m)	Percentages Fines	Sand	Gravel
1.1 − 2.1	31	35	34
2.1 − 3.1	6	26	68
3.1 − 4.1	12	34	54
4.1 − 5.1	14	29	57
5.1 − 5.8	12	36	52

SU 77 NE 19 | 7664 7775 | Near The Warren, Shiplake | Block D

Surface level + 34.0 m (+ 112 ft)
Water struck at surface
Shell and auger (modified) 6 in (152 mm) diameter
December 1972

Overburden 3.1 m (10.0 ft)
Mineral 7.3 m (24.0 ft)
Bedrock 0.6 m+ (2.0 ft+)

LOG

Geological Classification	Lithology	Thickness m	Thickness ft	Depth m	Depth ft
	Made ground	0.1	(0.5)	0.1	(0.5)
Alluvium	Silt, soft, bluish grey, with some peat	3.0	(10.0)	3.1	(10.0)
Younger River Gravels (undifferentiated)	Gravel Gravel: fine to coarse with scattered cobbles, subangular to well rounded black, white and brown flints and limestone with some quartz and ironstone. Well rounded chalk pebbles predominate in lowest 0.7 m Sand: mainly coarse and medium, brown to light brown	7.3	(24.0)	10.4	(34.0)
Upper Chalk	Chalk	0.6+	(2.0+)	11.0	(36.0)

GRADING

Mean for deposit

	%	mm	%
Gravel	59	+16	26
		−16+4	33
Sand	36	−4+1	19
		−1+¼	15
		−¼+1/16	2
Fines	5	−1/16	5

Bulk samples

Depth below surface (m)	Percentages Fines	Sand	Gravel
*3.1 − 4.1	2	23	75
*4.1 − 5.1	5	36	59
*5.1 − 6.1	1	39	60
*6.1 − 7.1	4	37	59
*7.1 − 8.1	5	44	51
*8.1 − 9.1	4	39	57
*9.1 − 9.7	3	23	74
*9.7 − 10.4	22	39	39

COMPOSITION

Percentages (in 4.75 to 9.5 mm fractions)

Depth below surface (m)	Flint	Quartz	Ironstone	Chalk	Limestone
3.1 − 4.1	52	2	3	0	43
4.1 − 5.1	51	8	5	2	34
5.1 − 6.1	56	3	10	1	29
6.1 − 7.1	72	4	5	4	15
7.1 − 8.1	77	2	4	4	13
8.1 − 9.1	no information				
9.1 − 9.7	no information				
9.7 − 10.4	no information				
Mean	62	4	5	2	27

Surface level +34.0 m (+112 ft)
Water struck at +30.9 m
Shell and auger (modified) 6 in (152 mm) diameter
January 1973

Overburden 3.1 m (10.0 ft)
Mineral 4.1 m (13.5 ft)
Bedrock 0.5 m+ (1.5 ft+)

LOG

Geological Classification	Lithology	Thickness m	Thickness ft	Depth m	Depth ft
	Soil	0.3	(1.0)	0.3	(1.0)
Alluvium	Clay, light brown with abundant gastropods	1.7	(5.5)	2.0	(6.5)
	Clay, silty, light brown, varying to greyish green and black, numerous gastropods and lamellibranchs at base	1.1	(3.5)	3.1	(10.0)
Younger River Gravels (undifferentiated)	Gravel Gravel: fine to coarse, subangular to subrounded black and brown flint with limestone and some ironstone, chalk and subrounded to well rounded quartz Sand: medium and coarse quartz, flint and chalk, white to grey	4.1	(13.5)	7.2	(23.5)
Upper Chalk	Chalk	0.5+	(1.5+)	7.7	(25.5)

GRADING

Mean for deposit

	%	mm	%
Gravel	61	+16	34
		−16+4	27
Sand	30	−4+1	16
		$-1+\frac{1}{4}$	14
		$-\frac{1}{4}+1/16$	0
Fines	9	−1/16	9

Bulk samples

Depth below surface (m)	Percentages Fines	Sand	Gravel
*3.1 − 4.1	17	28	55
*4.1 − 5.1	11	37	52
*5.1 − 6.1	5	26	69
*6.1 − 7.2	2	28	70

COMPOSITION

Depth below surface (m)	Percentages (in 4.75 to 9.5 mm fractions) Flint	Quartz	Ironstone	Chalk	Limestone	Miscellaneous
3.1 − 4.1	no information					
4.1 − 5.1	55	8	7	3	26	1
5.1 − 6.1	no information					
6.1 − 7.2	57	7	5	2	29	0
Mean	55	8	6	3	27	1

SU 77 NE 21	7678 7521	Model Farm, Sonning	Block D

Surface level + 53.7 m (+ 176 ft)
Water not struck
Shell and auger (modified) 6 in (152 mm) diameter
January 1973

Overburden 0.4 m (1.5 ft)
Mineral 4.3 m (14.0 ft)
Bedrock 1.0 m + (3.5 ft +)

LOG

Geological Classification	Lithology	Thickness m	ft	Depth m	ft
	Soil	0.2	(0.5)	0.2	(0.5)
Younger River Gravels (undifferentiated)	Clay, sandy, light brown, scattered flint pebbles	0.2	(0.5)	0.4	(1.5)
	Gravel	4.3	(14.0)	4.7	(15.5)
	Gravel: fine to coarse becoming coarser with depth, subrounded to well rounded flint with some well rounded quartz and quartzite and occasional sandstone				
	Sand: medium with some coarse, brown				
Reading Beds	Clay, mottled red, reddish grey and brown	1.0 +	(3.5 +)	5.7	(18.5)

GRADING

Mean for deposit

	%	mm	%
Gravel	57	+16	30
		−16+4	27
Sand	35	−4+1	12
		$-1+\frac{1}{4}$	21
		$-\frac{1}{4}+1/16$	2
Fines	8	−1/16	8

Bulk samples

Depth below surface (m)	Percentages Fines	Sand	Gravel
0.4 − 1.4	16	35	49
1.4 − 2.4	10	29	61
2.4 − 3.4	6	40	54
3.4 − 4.7	2	36	62

SU 77 NE 22 | 7734 7963 | Lower Shiplake, Shiplake | Block C

Surface level +37. 2 m (+122 ft)
Water struck at +32.0 m
Shell and auger (modified) 6 in (152 mm) diameter
December 1972

Overburden 0.6 m (2.0 ft)
Mineral 3.0 m (10.0 ft)
Waste 1.9 m (6.0 ft)
Bedrock 0.5 m+ (1.5 ft+)

LOG

Geological Classification	Lithology	Thickness m	ft	Depth m	ft
	Soil and made ground	0.6	(2.0)	0.6	(2.0)
Younger River Gravels (undifferentiated)	'Very clayey' gravel Gravel: mainly coarse with some fine, black and brown flint, subangular to subrounded with occasional quartz and quartzite Sand: medium to fine with some coarse, brown	3.0	(10.0)	3.6	(12.0)
	Clay, sandy with occasional pebbles, brown	1.9	(6.0)	5.5	(18.0)
Upper Chalk	Chalk with flint nodules	0.5+	(1.5+)	6.0	(19.5)

GRADING

Mean for deposit

	%	mm	%
Gravel	39	+16	26
		−16+4	13
Sand	32	−4+1	5
		$-1+\frac{1}{4}$	17
		$-\frac{1}{4}+1/16$	10
Fines	29	−1/16	29

Bulk samples

Depth below surface (m)	Percentages Fines	Sand	Gravel
0.6 − 1.5	40	48	12
1.5 − 2.5	11	17	72
2.5 − 3.6	36	32	32

SU 77 NE 23	7752 7796	Borough Marsh, near Wargrave	Block D

Surface level + 35.0 m (+ 115 ft)
Water struck at + 32.4 m
Shell and auger (modified) 6 in (152 mm) diameter
December 1972

Overburden 1.6 m (5.5 ft)
Mineral 6.6 m (21.5 ft)
Bedrock 0.5 m+ (1.5 ft+)

LOG

Geological Classification	Lithology	Thickness m	Thickness ft	Depth m	Depth ft
	Soil	0.3	(1.0)	0.3	(1.0)
Alluvium	Clay, brown, sandy with scattered flint pebbles	1.3	(4.5)	1.6	(5.5)
Younger River Gravels (undifferentiated)	Gravel Gravel: fine to coarse with occasional cobbles, subangular to subrounded flint, with limestone and some quartz, quartzite and a trace of chalk and ironstone Sand: mainly coarse and medium	6.6	(21.5)	8.2	(27.0)
Upper Chalk	Chalk with flint nodules	0.5+	(1.5+)	8.7	(28.5)

GRADING

Mean for deposit

	%	mm	%
Gravel	55	+16	32
		−16+4	33
Sand	29	−4+1	13
		$-1+\frac{1}{4}$	10
		$-\frac{1}{4}+1/16$	6
Fines	6	−1/16	6

Bulk samples

Depth below surface (m)	Percentages Fines	Sand	Gravel
1.6 – 2.6	26	36	38
*2.6 – 3.6	14	36	50
*3.6 – 4.6	2	10	88
*4.6 – 5.6	1	18	81
*5.6 – 6.6	11	32	57
*6.6 – 7.6	6	30	64
*7.6 – 8.2	2	16	82

COMPOSITION

Depth below surface (m)	Percentages (in 4.75 to 9.5 mm fractions) Flint	Quartz	Ironstone	Chalk	Limestone
1.6 – 2.6	76	5	3	3	13
2.6 – 3.6	70	6	2	3	19
3.6 – 4.6	70	12	4	2	11
4.6 – 5.6	75	7	1	2	14
5.6 – 6.6	77	5	3	2	13
6.6 – 7.6	79	6	3	4	8
7.6 – 8.2	83	6	2	3	6
Mean	75	7	3	3	12

SU 77 NE 24 7791 7647 East of Charvil Farm, Charvil Block D

Surface level +35.0 m (+115 ft)
Water not struck
Shell and auger (modified) 6 in (152 mm) diameter
January 1973

Overburden 0.8 m (2.5 ft)
Mineral 6.3 m (20.5 ft)
Bedrock 0.5 m+ (1.5 ft+)

LOG

Geological Classification	Lithology	Thickness m	ft	Depth m	ft
	Made ground	0.8	(2.5)	0.8	(2.5)
Younger River Gravels (undifferentiated)	Gravel Gravel: mainly coarse with some fine with scattered cobbles, subangular to subrounded flint with well rounded quartz and quartzite and a trace of sandstone and chalk Sand: medium and coarse with fine, mainly quartz and flint	6.3	(20.5)	7.1	(23.5)
Upper Chalk	Chalk with flint nodules	0.5+	(1.5+)	7.6	(25.0)

GRADING

Mean for deposit

	%	mm	%
Gravel	60	+16	55
		−16+4	5
Sand	31	−4+1	11
		$-1+\frac{1}{4}$	16
		$-\frac{1}{4}+1/16$	4
Fines	9	−1/16	9

Bulk samples

Depth below surface (m)	Percentages Fines	Sand	Gravel
0.8 − 1.6	40	48	12
1.6 − 2.6	9	32	59
2.6 − 3.6	4	26	70
3.6 − 4.6	5	29	66
4.6 − 5.6	1	23	76
5.6 − 6.6	3	28	69
6.6 − 7.1	5	32	63

SU 77 NE 25 7712 7600 Near The Wee Waif, Charvil Block D

Surface level +38.9 m (+128 ft)
Water not struck
Shell and auger (modified) 6 in (152 mm) diameter
January 1973

Overburden 2.2 m (7.0 ft)
Mineral 2.6 m (8.5 ft)
Bedrock 0.5 m+ (1.5 ft+)

LOG

Geological Classification	Lithology	Thickness m	ft	Depth m	ft
	Soil	0.2	(0.5)	0.2	(0.5)
Younger River Gravels (undifferentiated)	Clay, with sand and gravel	2.0	(6.5)	2.2	(7.0)
	'Clayey' gravel Gravel: mainly coarse with some fine, rare cobbles, subangular to subrounded flint with some subangular to well rounded quartz and quartzite, gravel increasing in amount with depth Sand: mainly coarse and medium, clayey throughout	2.6	(8.5)	4.8	(15.5)
Upper Chalk	Chalk, soft, with flints	0.5+	(1.5+)	5.3	(17.5)

GRADING

Mean for deposit

	%	mm	%
Gravel	63	+16	42
		−16+4	21
Sand	20	−4+1	9
		$-1+\frac{1}{4}$	9
		$-\frac{1}{4}+1/16$	2
Fines	17	−1/16	17

Bulk samples

Depth below surface (m)	Percentages Fines	Sand	Gravel
2.2 – 3.2	14	14	72
3.2 – 4.2	14	23	63
4.2 – 4.8	28	23	49

SU 77 NE 26 | 7807 7906 | Near Lashbrook House, Shiplake | Block D

Surface level + 32.7 m (+ 107 ft)
Water struck at + 31.3 m
Shell and auger (modified) 6 in (152 mm) diameter
December 1972

Overburden 0.8 m (2.5 ft)
Mineral 5.9 m (19.5 ft)
Bedrock 0.5 m+ (1.5 ft+)

LOG

Geological Classification	Lithology	Thickness m	ft	Depth m	ft
	Soil	0.2	(0.5)	0.2	(0.5)
Alluvium	Clay, light brown, with scattered flint pebbles	0.6	(2.0)	0.8	(2.5)
Younger River Gravels (undifferentiated)	Gravel Gravel: fine to coarse with scattered cobbles, subangular to subrounded black and brown flints, with limestone and with a trace of chalk, ironstone and subrounded to well rounded quartz and quartzite Sand: medium to coarse, mainly quartz and flint, white	5.9	(19.5)	6.7	(22.0)
Upper Chalk	Chalk	0.5+	(1.5+)	7.2	(23.5)

GRADING

Mean for deposit

	%	mm	%
Gravel	67	+16	36
		−16+4	31
Sand	29	−4+1	11
		−1+¼	17
		−1+1/16	1
Fines	4	−1/16	4

Bulk samples

Depth below surface (m)	Percentages Fines	Sand	Gravel
*0.8 − 1.8	1	32	67
*1.8 − 2.8	4	33	63
*2.8 − 3.8	7	35	58
*3.8 − 4.8	3	6	91
*4.8 − 5.8	2	18	80
*5.8 − 6.7	5	49	46

COMPOSITION

Depth below surface (m)	Percentages (in 4.75 to 9.5 mm fractions) Flint	Quartz	Ironstone	Chalk	Limestone
0.8 − 1.8	75	4	5	2	14
1.8 − 2.8	68	3	3	2	24
2.8 − 3.8	55	5	8	2	30
3.8 − 4.8	76	3	4	2	15
4.8 − 5.8	70	5	5	3	19
5.8 − 6.7	84	2	3	4	7
Mean	71	4	4	3	18

SU 77 NE 27   7849 7728   Near Loddon Park Farm, Twyford   Block E

Surface level + 39.0 m (+ 128 ft)
Water not struck
Shell and auger (modified) 6 in (152 mm) diameter
January 1973

Overburden 2.0 m (6.5 ft)
Mineral 1.0 m (3.5 ft)
Waste 0.6 m (2.0 ft)
Bedrock 0.4 m+ (1.5 ft+)

LOG

Geological Classification	Lithology	Thickness m	ft	Depth m	ft
	Soil and subsoil	0.5	(1.5)	0.5	(1.5)
	? Made ground	1.5	(5.0)	2.0	(6.5)
Younger River Gravels (undifferentiated)	'Very clayey' gravel Gravel: fine to coarse subangular to subrounded flint with some subrounded to well rounded quartz and quartzite Sand: medium and coarse, clayey brown	1.0	(3.5)	3.0	(10.0)
	Clay, with sand and gravel	0.6	(2.0)	3.6	(12.0)
Upper Chalk	Chalk with flint	0.4+	(1.5+)	4.0	(13.0)

GRADING

Mean for deposit

	%	mm	%
Gravel	50	+16	22
		−16+4	28
Sand	20	−4+1	9
		−1+4	10
		−¼+1/16	1
Fines	30	−1/16	30

Bulk samples

Depth below surface (m)	Percentages Fines	Sand	Gravel
2.0 − 3.0	30	20	50

SU 77 SW 8	7044 7475	Thames Side Promenade, Reading	Block D

Surface level (+ 39.0 m) + 128 ft
Water struck at + 35.2 m
Shell and auger (modified) 6 in (152 mm) diameter
November 1972

Overburden 3.8 (12.5 ft)
Mineral 6.9 m (22.5 ft)
Bedrock 0.5 m + (1.5 ft+)

LOG

Geological Classification	Lithology	Thickness m	Thickness ft	Depth m	Depth ft
	Made ground	2.2	(7.0)	2.2	(7.0)
Alluvium	Clay, silty, greyish brown	1.6	(5.0)	3.8	(12.5)
Younger River Gravels (undifferentiated)	Gravel Gravel: fine to coarse, some cobbles, subangular to well rounded, brown and black flint, with some limestone, well rounded quartz and quartzite, fine grained sandstone and a trace of ironstone Sand: medium to coarse, quartz and flint	6.9	(22.5)	10.7	(35.0)
Upper Chalk	Chalk	0.5+	(1.5+)	11.2	(36.5)

GRADING

Mean for deposit

	%	mm	%
Gravel	62	+16	37
		−16+4	25
Sand	34	−4+1	15
		$-1+\frac{1}{4}$	17
		$-\frac{1}{4}+1/16$	2
Fines	4	−1/16	4

Bulk samples

Depth below surface (m)	Percentages Fines	Sand	Gravel
*3.8 − 4.8	6	44	50
*4.8 − 5.8	2	26	72
*5.8 − 6.8	1	28	71
*6.8 − 7.8	1	30	69
*7.8 − 8.8	1	29	70
*8.8 − 9.8	19	58	23
*9.8 −10.7	1	21	78

SU 77 SW 9 7037 7431 Cow Lane, Reading Block D

Surface level (+ 37.8 m) + 124 ft
Water struck at + 36.5 m
Shell and auger (modified) 6 in (152 mm) diameter
November 1972

Overburden 1.3 m (4.5 ft)
Mineral 6.8 m (22.5 ft)
Bedrock 0.5 m + (1.5 ft +)

LOG

Geological Classification	Lithology	Thickness m	Thickness ft	Depth m	Depth ft
	Soil	0.1	(0.5)	0.1	(0.5)
Younger River Gravels (undifferentiated)	Clay, sandy, brown, trace of gravel	1.2	(4.0)	1.3	(4.5)
	Gravel	6.8	(22.5)	8.1	(26.5)
	Gravel: mainly coarse with some fine, subangular to well rounded flint with limestone and chalk and with some ironstone and well rounded quartz and quartzite				
	Sand: mainly medium and coarse flint and quartz, pale grey				
Upper Chalk	Chalk	0.5+	(1.5+)	8.6	(28.0)

GRADING

Mean for deposit

	%	mm	%
Gravel	67	+16	43
		−16+4	24
Sand	27	−4+1	10
		$-1+\frac{1}{4}$	15
		$-\frac{1}{4}+1/16$	2
Fines	6	−1/16	6

Bulk samples

Depth below surface (m)	Percentages Fines	Sand	Gravel
*1.3 − 2.3	13	49	38
*2.3 − 3.3	8	21	71
*3.3 − 4.3	5	26	69
*4.3 − 5.3	1	17	82
*5.3 − 6.3	8	25	67
*6.3 − 7.3	0	29	71
*7.3 − 8.1	4	24	72

COMPOSITION

Depth below surface (m)	Percentages (in 4.75 to 9.5 mm fractions) Flint	Quartz	Ironstone	Chalk	Limestone
1.3 − 2.3	63	3	3	3	28
2.3 − 3.3	60	5	3	6	26
3.3 − 4.3	57	6	4	6	27
4.3 − 5.3	74	2	4	5	15
5.3 − 6.3	74	2	4	4	16
6.3 − 7.3	71	2	3	5	19
7.3 − 8.1	69	1	2	6	22
Mean	67	3	3	5	22

SU 77 SW 10 7080 7104 Fobney Meadow, Reading Block D

Surface level (+ 39.0 m) + 128 ft
Water struck at + 35.0 m
Shell and auger (modified) 6 in (152 mm) diameter
December 1972

Overburden 4.0 (13.0 ft)
Mineral 3.3 m (11.0 ft)
Bedrock 1.0 m+ (3.5 ft +)

LOG

Geological Classification	Lithology	Thickness m	Thickness ft	Depth m	Depth ft
	Soil	0.3	(1.0)	0.3	(1.0)
Alluvium	Clay, sandy and silty, light grey passing into greyish blue	3.2	(10.5)	3.5	(11.5)
	Peat and clay, greyish blue	0.5	(1.5)	4.0	(13.0)
Younger River Gravels (undifferentiated)	Gravel Gravel: mainly fine with some coarse, subangular to subrounded flint with a few cobbles, with some quartz and quartzite Sand: medium with fine, grey to dark grey	3.3	(11.0)	7.3	(24.0)
Reading Beds	Clay, mottled grey-blue	1.0+	(3.5+)	8.3	(27.0)

GRADING

Mean for deposit

	%	mm	%
Gravel	66	+16	16
		−16+4	50
Sand	29	−4+1	10
		−1+¼	18
		−¼+1/16	1
Fines	5	−1/16	5

Bulk samples

Depth below surface (m)	Percentages Fines	Sand	Gravel
*4.0 − 5.0	4	28	68
*5.0 − 6.0	3	29	68
*6.0 − 7.3	8	29	63

SU 77 SW 11 | 7029 7035 | West of Manor Farm, Reading | Block D

Surface level (+ 38.1) + 125 ft
Water struck at + 36.4 m
Shell and auger (modified) 6 in (152 mm) diameter
December 1972

Overburden 1.1 m (3.5 ft)
Mineral 3.9 (13.0 ft)
Bedrock 6.7 m+ (22.0 ft+)

LOG

Geological Classification	Lithology	Thickness m	ft	Depth m	ft
	Soil	0.2	(0.5)	0.2	(0.5)
Younger River Gravels (undifferentiated)	Clay, silty and sandy, light brown, scattered pebbles	0.9	(3.0)	1.1	(3.5)
	Gravel Gravel: mainly coarse with some fine, predominantly flint with some quartz and quartzite Sand: medium with some coarse, very little clay, grey to dark grey	3.9	(13.0)	5.0	(16.5)
	Sand, fine, very silty, light grey	6.7 +	(22.0 +)	11.7	(38.5)
Reading Beds	Clay, silty, grey-green				

GRADING

Mean for deposit

	%	mm	%
Gravel	73	+16	26
		−16+4	47
Sand	24	−4+1	8
		−1+¼	14
		−¼+1/16	2
Fines	3	−1/16	3

Bulk samples

Depth below surface (m)	Percentages Fines	Sand	Gravel
*1.1 − 2.1	6	26	68
*2.1 − 3.1	2	30	68
*3.1 − 4.1	1	20	79
*4.1 − 5.0	1	20	79

SU 77 SW 12 7236 7402 King's Meadow, Reading

Surface level (+ 35.7 m) + 117 ft
Water struck at + 34.4 m
Shell and auger (modified) 6 in (152 mm) diameter
December 1972

Overburden 1.3 m (4.5 ft)
Mineral 8.8 m (29.0 ft)
Bedrock 0.5 m+ (1.5 ft+)

LOG

Geological Classification	Lithology	Thickness m	ft	Depth m	ft
	Made ground	0.4	(1.5)	0.4	(1.5)
Alluvium	Clay, silty and sandy in parts, scattered rounded flint and quartz pebbles, light brown to yellowish brown	0.9	(3.0)	1.3	(4.5)
Younger River Gravels (undifferentiated)	Gravel Gravel: fine to coarse with some cobbles, subangular to well rounded, mainly well rounded, black and occasionally brown flint and limestone with well rounded quartz and quartzite, ironstone and some chalk Sand: medium and coarse, flint and quartz with little clay	8.8	(29.0)	10.1	(33.0)
Upper Chalk	Chalk	0.5+	(1.5+)	10.6	(35.0)

GRADING

Mean for deposit

	%	mm	%
Gravel	59	+16	32
		−16+4	27
Sand	37	− 4+1	18
		$-1+\frac{1}{4}$	17
		$-\frac{1}{4}+1/16$	2
Fines	4	−1/16	4

Bulk samples

Depth below surface (m)	Percentages Fines	Sand	Gravel
*1.3 − 2.3	10	39	51
*2.3 − 3.3	0	33	67
*3.3 − 4.3	3	35	62
*4.3 − 5.3	8	46	46
*5.3 − 6.3	3	46	51
*6.3 − 7.3	3	21	76
*7.3 − 8.3	2	38	60
*8.3 − 9.3	3	36	61
*9.3 −10.1	4	41	55

COMPOSITION

Depth below surface (m)	Percentages (in 4.75 to 9.5 mm fractions) Flint	Quartz	Ironstone	Chalk	Limestone	Miscellaneous
1.3 − 2.3	45	6	5	1	42	1
2.3 − 3.3	47	10	5	0	34	4
3.3 − 4.3	40	9	7	1	41	2
4.3 − 5.3	56	3	8	1	30	2
5.3 − 6.3	48	3	8	3	35	3
6.3 − 7.3	43	5	9	3	33	7
7.3 − 8.3	40	7	10	2	31	10
8.3 − 9.3	42	13	11	0	29	5
9.3 −10.1	61	4	7	4	21	3
Mean	47	7	8	2	32	4

SU 77 SW 13 | 7353 7487 | Near Lowfield Farm, East of Caversham | Block D

Surface level (+ 36.9 m) + 121 ft
Water struck at + 35.3 m
Shell and auger (modified) 6 in (152 mm) diameter
December 1972

Overburden 0.7 m (2.5 ft)
Mineral 7.7 m (25.5 ft)
Bedrock 0.5 m+ (1.5 ft+)

LOG

Geological Classification	Lithology	Thickness m	Thickness ft	Depth m	Depth ft
	Soil	0.3	(1.0)	0.3	(1.0)
Younger River Gravels (undifferentiated)	Clay, sandy, scattered pebbles, light brown	0.4	(1.5)	0.7	(2.5)
	Gravel Gravel: fine to coarse, brown and black flint with limestone and with some quartz, quartzite, ironstone and chalk Sand: mainly medium and coarse, flint and quartz, light brown	7.7	(25.5)	8.4	(27.5)
Upper Chalk	Chalk	0.5+	(1.5+)	8.9	(29.0)

GRADING

Mean for deposit

	%	mm	%
Gravel	66	+16	35
		−16+4	31
Sand	29	−4+1	14
		−1+¼	13
		−¼+1/16	2
Fines	5	−1/16	5

Bulk samples

Depth below surface (m)	Percentages Fines	Sand	Gravel
0.7 − 1.7	22	50	28
*1.7 − 2.7	2	33	65
*2.7 − 3.7	3	34	63
*3.7 − 4.7	1	25	74
*4.7 − 5.7	1	5	94
*5.7 − 6.7	3	34	63
*6.7 − 7.7	2	20	78
*7.7 − 8.4	5	33	62

COMPOSITION

Depth below surface (m)	Percentages (in 4.75 to 9.5 mm fractions) Flint	Quartz	Ironstone	Chalk	Limestone	Miscellaneous
0.7 − 1.7	63	3	2	11	18	0
1.7 − 2.7	No sample					
2.7 − 3.7	No sample					
3.7 − 4.7	71	3	3	6	14	3
4.7 − 5.7	No sample					
5.7 − 6.7	72	4	5	2	16	1
6.7 − 7.7	77	2	3	2	15	1
7.7 − 8.4	61	11	5	9	13	1
Mean	71	4	3	6	15	1

Surface level (+36.3 m) +119 ft
Water struck at +35.3 m
Shell and auger (modified) 6 in (152 mm) diameter
December 1972

Overburden 1.0 m (3.5 ft)
Mineral 8.3 m (27.0 ft)
Bedrock 0.5 m+ (1.5 ft+)

LOG

Geological Classification	Lithology	Thickness m	Thickness ft	Depth m	Depth ft
	Soil	0.2	(0.5)	0.2	(0.5)
Alluvium	Clay, silty, with some flint gravel, light brown	0.8	(2.5)	1.0	(3.5)
Younger River Gravels (undifferentiated)	Gravel, with clay from 5.6 to 5.7 m Gravel: fine with coarse and occasional cobbles, subangular to subrounded flint with limestone and with some ironstone, chalk and subangular to well rounded quartz and quartzite Sand: mainly medium with some coarse, light brown	8.3	(27.0)	9.3	(30.5)
Upper Chalk	Chalk	0.5+	(1.5+)	9.8	(32.0)

GRADING

Mean for deposit

	%	mm	%
Gravel	51	+16	21
		−16+4	30
Sand	43	−4+1	17
		−1+¼	24
		−¼+1/16	2
Fines	6	−1/16	6

Bulk samples

Depth below surface (m)	Percentages Fines	Sand	Gravel
*1.0 − 2.0	12	33	55
*2.0 − 3.0	6	39	55
*3.0 − 4.0	1	45	54
*4.0 − 5.0	1	57	42
*5.0 − 5.6	7	69	24
*5.7 − 6.7	7	45	48
*6.7 − 7.7	8	43	49
*7.7 − 8.7	3	32	65
*8.7 − 9.3	11	33	56

COMPOSITION

Depth below surface (m)	Percentages (in 4.75 to 9.5 mm fractions) Flint	Quartz	Ironstone	Chalk	Limestone	Miscellaneous
1.0 − 2.0	96	1	0	0	0	3
2.0 − 3.0	No sample					
3.0 − 4.0	53	8	7	2	29	1
4.0 − 5.0	76	7	1	4	11	1
5.0 − 5.6	No sample					
5.7 − 6.7	82	2	2	5	9	0
6.7 − 7.7	76	4	2	5	13	0
7.7 − 8.7	68	2	6	5	19	0
8.7 − 9.3	68	3	6	2	21	0
Mean	74	4	3	3	15	1

SU 77 SE 41 — 7582 7465 — East of Holme Park Farm, Sonning

Surface level (+ 60.4 m) + 198 ft
Water not struck
Shell and auger (modified) 6 in (152 mm) diameter
January 1973

Overburden 1.0 m (3.5 ft)
Mineral 3.3 m (11.0 ft)
Bedrock 1.4 m+ (4.5 ft+)

LOG

Geological Classification	Lithology	Thickness m	Thickness ft	Depth m	Depth ft
	Soil	0.2	(0.5)	0.2	(0.5)
Younger River Gravels (undifferentiated)	Clay, silty, with some fine to medium flint gravel	0.8	(2.5)	1.0	(3.5)
	'Clayey' gravel Gravel: fine with some coarse, subangular to subrounded flint with some subrounded to well rounded quartz and quartzite. Sand: slightly clayey, becoming more clayey with depth, brown to yellow brown	3.3	(11.0)	4.3	(14.0)
Reading Beds	Clay, silty and slightly sandy at first, mottled grey, brown and red	1.4+	(4.5+)	5.7	(18.5)

GRADING

Mean for deposit

	%	mm	%
Gravel	51	+16	18
		−16+4	33
Sand	32	−4+1	13
		−1+¼	17
		−¼+1/16	2
Fines	17	−1/16	17

Bulk samples

Depth below surface (m)	Percentages Fines	Sand	Gravel
1.0 − 2.0	11	32	57
2.0 − 3.0	16	35	49
3.0 − 4.3	22	31	47

SU 77 SE 42 7673 7424 Butts Hill, Woodley and Sandford

Surface level (+ 52.1 m) + 171 ft
Water struck at + 49.8 m
Shell and auger (modified) 6 in (152 mm) diameter
January 1973

Overburden 0.4 m (1.5 ft)
Mineral 2.8 m (9.0 ft)
Bedrock 2.8 m + (9.0 ft +)

LOG

Geological Classification	Lithology	Thickness m	ft	Depth m	ft
	Soil	0.4	(1.5)	0.4	(1.5)
Younger River Gravels (undifferentiated)	'Clayey' sandy gravel Gravel: fine and coarse with occasional cobbles, subrounded to well rounded flint with some quartz and a trace of quartzite Sand: medium with some coarse, slightly clayey throughout, yellow brown	2.8	(9.0)	3.2	(10.5)
Reading Beds	Clay, orange brown becoming dark grey at 4.8 m, soft, silty with occasional slightly sandy horizons	2.8 +	(9.0 +)	6.0	(19.5)

GRADING

Mean for deposit

	%	mm	%
Gravel	39	+16	16
		−16+4	23
Sand	46	−4+1	11
		$-1+\frac{1}{4}$	33
		$-\frac{1}{4}+1/16$	2
Fines	15	−1/16	15

Bulk samples

Depth below surface (m)	Percentages Fines	Sand	Gravel
0.4 – 1.4	17	38	45
1.4 – 2.4	12	56	32
*2.4 – 3.2	17	44	39

SU 77 SE 43 7665 7155 Loddon Bridge, Winnersh Block E

Surface level (+ 38.1 m) + 125 ft
Water struck at + 36.7 m
Shell and auger (modified) 6 in (152 mm) diameter
January 1973

Overburden 1.0 m (3.5 ft)
Mineral 6.6 m (21.5 ft)
Bedrock 0.5 m + (1.5 ft +)

LOG

Geological Classification	Lithology	Thickness m	ft	Depth m	ft
	Soil	0.3	(1.0)	0.3	(1.0)
Alluvium	Clay, sandy, scattered pebbles, light brown	0.7	(2.5)	1.0	(3.5)
Younger River Gravels (undifferentiated)	Gravel Gravel: fine and coarse, subangular to subrounded flint with some subrounded to rounded quartz and a trace of quartzite Sand: mainly medium with some coarse, brown, fine, clayey and almost gravel free from 1.0 m to 2.0 m	6.6	(21.5)	7.6	(25.0)
London Clay	Clay, dark grey	0.5+	(1.5+)	8.1	(26.5)

GRADING

Mean for deposit

	%	mm	%
Gravel	64	+16	29
		−16+4	35
Sand	30	−4+1	9
		$-1+\frac{1}{4}$	18
		$-\frac{1}{4}+1/16$	3
Fines	6	−1/16	6

Bulk samples

Depth below surface (m)	Percentages Fines	Sand	Gravel
*1.0 − 2.0	22	70	8
*2.0 − 3.0	5	25	70
*3.0 − 4.0	1	22	77
*4.0 − 5.0	2	21	77
*5.0 − 6.0	6	17	77
*6.0 − 7.0	2	25	73
*7.0 − 7.6	5	27	68

Surface level (+ 39.6 m) + 130 ft
Water struck at + 38.6 m
Shell and auger (modified) 6 in (152 mm) diameter
January 1973

Overburden 0.5 m (1.5 ft)
Mineral 3.4 m (11.0 ft)
Bedrock 0.5 m+ (1.5 ft+)

LOG

Geological Classification	Lithology	Thickness m	ft	Depth m	ft
	Soil	0.2	(0.5)	0.2	(0.5)
Younger River Gravels (undifferentiated)	Clay, sandy, scattered flint pebbles, brown	0.3	(1.0)	0.5	(1.5)
	'Clayey' gravel Gravel: fine to coarse subangular to well rounded flints with some quartz and quartzite Sand: mainly medium with some coarse and fine, clayey throughout, brown to light brown	3.4	(11.0)	3.9	(13.0)
London Clay	Clay, blue-grey	0.5+	(1.5+)	4.4	(14.5)

GRADING

Mean for deposit

	%	mm	%
Gravel	48	+16	20
		−16+4	28
Sand	34	−4+1	8
		$-1+\frac{1}{4}$	22
		$-\frac{1}{4}+1/16$	4
Fines	18	−1/16	18

Bulk samples

Depth below surface (m)	Percentages Fines	Sand	Gravel
0.5 − 1.1	25	30	45
*1.1 − 2.1	19	38	43
*2.1 − 3.1	15	24	61
*3.1 − 3.9	17	41	42

SU 77 SE 45 7664 7029 Sindlesham Mill, West of Winnersh Block E

Surface level (+38.4 m) +126 ft
Water struck at +36.4 m
Shell and auger (modified) 6 in (152 mm) diameter
January 1973

Overburden 1.0 (3.5 ft)
Mineral 3.1 m (10.0 ft)
Bedrock 0.5 m+ (1.5 ft+)

LOG

Geological Classification	Lithology	Thickness m	ft	Depth m	ft
	Soil	0.2	(0.5)	0.2	(0.5)
Alluvium	Clay, silty and sandy, brown	0.8	(2.5)	1.0	(3.5)
Younger River Gravels (undifferentiated)	'Clayey' gravel Gravel: fine with some coarse, subrounded to well rounded, mainly flint with a trace of quartz and quartzite Sand: medium with some coarse, clayey, but becoming less clayey with depth, brown with grey and black streaks	3.1	(10.0)	4.1	(13.5)
London Clay	Clay, blue-grey	0.5+	(1.5+)	4.6	(15.0)

GRADING

Mean for deposit

	%	mm	%
Gravel	45	+16	14
		−16+4	31
Sand	40	−4+1	11
		$-1+\frac{1}{4}$	26
		$-\frac{1}{4}+1/16$	3
Fines	15	−1/16	15

Bulk samples

Depth below surface (m)	Percentages Fines	Sand	Gravel
1.0 − 2.0	19	38	43
*2.0 − 3.0	16	25	59
*3.0 − 4.1	11	55	34

SU 77 SE 46 7707 7127 Near Loddonbridge Farm, Winnersh Block E

Surface level (+ 39.3 m) + 129 ft
Water struck at + 37.8 m
Shell and auger (modified) 6 in (152 mm) diameter
January 1973

Overburden 0.8 m (2.5 ft)
Mineral 3.9 m (13.0 ft)
Bedrock 0.7 m+ (2.5 ft+)

LOG

Geological Classification	Lithology	Thickness m	Thickness ft	Depth m	Depth ft
	Soil	0.2	(0.5)	0.2	(0.5)
Younger River Gravels (undifferentiated)	Clay, sandy in parts, brown	0.6	(2.0)	0.8	(2.5)
	Gravel Gravel: fine to coarse, mainly flint with some quartz and a trace of quartzite, subangular to well rounded Sand: medium to coarse, clayey near top	3.9	(13.0)	4.7	(15.5)
London Clay	Clay, grey brown, slightly silty, becoming less silty with depth	0.7+	(2.5+)	5.4	(17.5)

GRADING

Mean for deposit

	%	mm	%
Gravel	61	+16	27
		−16+4	34
Sand	30	−4+1	14
		$-1+\frac{1}{4}$	14
		$-\frac{1}{4}+1/16$	2
Fines	9	−1/16	9

Bulk samples

Depth below surface (m)	Percentages Fines	Sand	Gravel
*0.8 − 1.8	19	43	38
*1.8 − 2.8	1	19	80
*2.8 − 3.8	1	27	72
*3.8 − 4.7	13	32	55

SU 77 SE 47 | 7853 7432 | Near Whistley Bridge, Whistley Green | Block E

Surface level (+ 36.0 m) + 118 ft
Water struck at + 33.7 m
Shell and auger (modified) 6 in (152 mm) diameter
January 1973

Overburden 1.9 m (6.0 ft)
Mineral 1.8 m (6.0 ft)
Bedrock 14.6 m + (48.0 ft +)

LOG

Geological Classification	Lithology	Thickness m	Thickness ft	Depth m	Depth ft
Alluvium	Clay, silty and peaty, black and yellowish brown	1.9	(6.0)	1.9	(6.0)
Younger River Gravels (undifferentiated)	Gravel Gravel: fine to coarse, subangular to subrounded black and brown flints, with rounded quartz and quartzite Sand: mainly coarse with some medium, flint with quartz, white to pale brown	1.8	(6.0)	3.7	(12.0)
Reading Beds	Clay, slightly silty in parts, mottled grey and red	14.6+	(48.0 +)	18.3	(60.0)

GRADING

Mean for deposit

	%	mm	%
Gravel	74	+16	32
		−16 + 4	42
Sand	18	−4 + 1	11
		$-1+\frac{1}{4}$	6
		$-\frac{1}{4}+1/16$	1
Fines	8	−1/16	8

Bulk samples

Depth below surface (m)	Percentages Fines	Sand	Gravel
*1.9 − 2.9	8	11	81
*2.9 − 3.7	8	32	60

SU 77 SE 48	7882 7244	Near Jolly Farmer, Hurst	Block E

Surface level (+ 38.4 m) + 126 ft
Water struck at + 36.7 m
Shell and auger (modified) 6 in (152 mm) diameter
January 1973

Overburden 0.4 m (1.5 ft)
Mineral 3.0 (10.0 ft)
Bedrock 0.5 m+ (1.5 ft+)

LOG

Geological Classification	Lithology	Thickness m	Thickness ft	Depth m	Depth ft
	Soil	0.4	(1.5)	0.4	(1.5)
Younger River Gravels (undifferentiated)	'Clayey' gravel Gravel: fine to coarse, brown and black subangular to subrounded flints with some subrounded to well rounded quartz and quartzite Sand: mainly medium with some coarse, brown to light brown	3.0	(10.0)	3.4	(11.0)
London Clay	Clay, smooth, black to dark grey	0.5+	(1.5+)	3.9	(13.0)

GRADING

Mean for deposit

	%	mm	%
Gravel	51	+16	22
		−16+4	29
Sand	31	−4+1	9
		−1+$\frac{1}{4}$	19
		−$\frac{1}{4}$+1/16	3
Fines	18	−1/16	18

Bulk samples

Depth below surface (m)	Percentages Fines	Sand	Gravel
0.4 − 1.4	37	28	35
*1.4 − 2.4	11	31	58
*2.4 − 3.4	5	35	60

SU 77 SE 49 7950 7430 Whistley Green, Hurst Block E

Surface level (+ 36.6 m) + 120 ft
Water not struck
Shell and auger (modified) 6 in (152 mm) diameter
January 1973

Overburden 2.2 m (7.0 ft)
Mineral 1.3 m (4.5 ft)
Bedrock 0.5 m+ (1.5 ft+)

LOG

Geological Classification	Lithology	Thickness m	Thickness ft	Depth m	Depth ft
	Soil	0.2	(0.5)	0.2	(0.5)
Younger River Gravels (undifferentiated)	Clay, sandy, yellowish brown, scattered pebbles	1.1	(3.5)	1.3	(4.5)
	Clay, silty and sandy, with some flint gravel	0.9	(3.0)	2.2	(7.0)
	'Clayey' gravel Gravel: mainly fine with some coarse, subangular to well rounded brown flint with some quartz and quartzite Sand: medium with some coarse, fairly clayey, brown	1.3	(4.5)	3.5	(11.5)
Reading Beds	Clay, mottled red and grey	0.5+	(1.5+)	4.0	(13.0)

GRADING

Mean for deposit

	%	mm	%
Gravel	48	+16	10
		−16+4	38
Sand	38	−4+1	14
		$-1+\frac{1}{4}$	22
		$-\frac{1}{4}+1/16$	2
Fines	14	−1/16	14

Bulk samples

Depth below surface (m)	Percentages Fines	Sand	Gravel
2.2 – 3.5	14	38	48

SU 77 SE 50 | 7915 7361 | Whistley Park, Hurst | Block E

Surface level (+37.2 m) +122 ft
Water struck at +36.4 m
Shell and auger (modified) 6 in (152 mm) diameter
January 1973

Overburden 0.3 m (1.0 ft)
Mineral 3.3 m (11.0 ft)
Bedrock 0.5 m+ (1.5 ft+)

LOG

Geological Classification	Lithology	Thickness m	Thickness ft	Depth m	Depth ft
	Soil	0.3	(1.0)	0.3	(1.0)
Younger River Gravels (undifferentiated)	Gravel Gravel: fine with some coarse subangular to subrounded flint with some fine, well rounded quartz and a trace of quartzite Sand: medium with coarse, quartz and flint, pale brown	3.3	(11.0)	3.6	(12.0)
Reading Beds	Clay, mottled red and grey	0.5+	(1.5+)	4.4	(13.5)

GRADING

Mean for deposit

	%	mm	%
Gravel	60	+16	22
		−16+4	38
Sand	31	−4+1	11
		$-1+\frac{1}{4}$	18
		$-\frac{1}{4}+1/16$	2
Fines	9	−1/16	9

Bulk samples

Depth below surface (m)	Percentages Fines	Sand	Gravel
*0.3 − 1.3	14	32	54
*1.3 − 2.3	12	40	48
*2.3 − 3.6	4	22	74

SU 77 SE 51	7944 7181	Near The Furze, Hurst	Block E

Surface level (+ 39.2 m) + 129 ft
Water struck at + 38.2 m
Shell and auger (modified) 6 in (152 mm) diameter
January 1973

Overburden 0.5 m (1.5 ft)
Mineral 2.1 m (7.0 ft)
Bedrock 0.5 m+ (1.5 ft+)

LOG

Geological Classification	Lithology	Thickness m	Thickness ft	Depth m	Depth ft
	Soil and made ground	0.5	(1.5)	0.5	(1.5)
Younger River Gravels (undifferentiated)	Gravel Gravel: mainly fine with a little coarse, subangular to subrounded with some well rounded, mainly brown flint with some quartz and quartzite Sand: medium and coarse, fairly clayey throughout, brown to light brown	2.1	(7.0)	2.6	(8.5)
London Clay	Clay, brown to grey brown	0.5+	(1.5+)	3.1	(10.0)

GRADING

Mean for deposit

	%	mm	%
Gravel	54	+16	8
		−16+4	46
Sand	33	−4+1	19
		$-1+\frac{1}{4}$	13
		$-\frac{1}{4}+1/16$	1
Fines	13	−1/16	13

Bulk samples

Depth below surface (m)	Percentages Fines	Sand	Gravel
*0.5 − 1.5	15	32	53
*1.5 − 2.6	11	35	54

SU 78 NW 8 7054 8645 Lower Common Wood, Nettlebed

Surface level (+ 175.0 m) + 574 ft
Water not struck
Shell and auger (modified) 6 in (152 mm) diameter
July 1972

Waste 1.7 m (5.5 ft)
Bedrock 0.5 m+ (1.5 ft+)

LOG

Geological Classification	Lithology	Thickness m	ft	Depth m	ft
	Soil	0.1	(0.5)	0.1	(0.5)
Clay-with-flints	Clay, sandy, light brown, with scattered large flints	0.5	(1.5)	0.6	(2.0)
	Clay, stiff, orange and reddish brown, with large irregularly shaped flints	1.1	(3.5)	1.7	(5.5)
Upper Chalk	Chalk, white, streaked reddish brown	0.5+	(1.5+)	2.2	(7.0)

SU 78 NW 9 7054 8534 Highmoor Common Wood, Highmoor

Surface level (+ 155.1 m) + 509 ft
Water not struck
Shell and auger (modified) 6 in (152 mm) diameter
July 1972

Waste 6.4 m (21.0 ft)
Bedrock 0.8 m+ (2.5 ft+)

LOG

Geological Classification	Lithology	Thickness m	ft	Depth m	ft
	Soil	0.3	(1.0)	0.3	(1.0)
Older River Gravels (undifferentiated)	Gravel, mainly flint, with silt, sand and clay	0.8	(2.5)	1.1	(3.5)
	Clay, silty, reddish brown, with nodular flints	5.3	(17.5)	6.4	(21.0)
Upper Chalk	Chalk, with bands of brown clay	0.8+	(2.5+)	7.2	(23.5)

SU 78 NW 10 7160 8871 Oak Farm, near Maidensgrove

Surface level (+190.8 m) +626 ft
Water not struck
Shell and auger (modified) 6 in (152 mm) diameter
November 1972

Waste 6.8 m (22.5 ft)
Bedrock 0.5 m+ (1.5 ft+)

Geological Classification	LOG Lithology	Thickness m	ft	Depth m	ft
	Soil	0.3	(1.0)	0.3	(1.0)
Clay-with-flints	Clay, dark brown, sandy, with flints	0.2	(0.5)	0.5	(1.5)
	Clay, orange brown, sandy, with coarse to cobble-sized, subrounded to sub-angular flint with some quartz and quartzite	2.0	(6.5)	2.5	(8.0)
	Flint, mainly cobble-sized nodules, with light brown clay	4.3	(14.0)	6.8	(22.5)
Upper Chalk	Chalk, with flints	0.5+	(1.5+)	7.3	(24.0)

SU 78 NW 11 7163 8596 Nettlebed Woods, near Nettlebed

Surface level (+159.1 m) +522 ft
Water not struck
Shell and auger (modified) 6 in (152 mm) diameter
August 1972

Waste 7.9 m (26.0 ft)
Bedrock 0.5 m+ (1.5 ft+)

Geological Classification	LOG Lithology	Thickness m	ft	Depth m	ft
	Soil	0.3	(1.0)	0.3	(1.0)
Clay-with-flints	Clay, flint gravel and sand	1.8	(6.0)	2.1	(7.0)
	Clay, reddish brown, with coarse and cobble-sized flints	5.8	(19.0)	7.9	(26.0)
Upper Chalk	Chalk	0.5+	(1.5+)	8.4	(27.5)

SU 78 NW 12 7248 8552 Bushycopse Shaw, Bix

Surface level (+130.5 m) +428 ft
Water not struck
Shell and auger (modified) 6 in (152 mm) diameter
August 1972

Waste 1.0 m (3.5 ft)
Bedrock 4.0 m+ (13.0 ft+)

LOG

Geological Classification	Lithology	Thickness m	Thickness ft	Depth m	Depth ft
	Soil	0.1	(0.5)	0.1	(0.5)
Older River Gravels (undifferentiated)	Gravel, with clay, sand and silt	0.9	(3.0)	1.0	(3.5)
Reading Beds	Clay, silty, reddish brown, mottled grey	4.0+	(13.0+)	5.0	(16.5)

SU 78 NW 13 7493 8583 Crockmore Farm, Fawley

Surface level (+135.6 m) +445 ft
Water not struck
Shell and auger (modified) 6 in (152 mm) diameter
November 1972

Waste 3.3 m (11.0 ft)
Bedrock 0.5 m+ (1.5 ft+)

LOG

Geological Classification	Lithology	Thickness m	Thickness ft	Depth m	Depth ft
	Soil	0.1	(0.5)	0.1	(0.5)
Clay-with-flints	Clay and silt with sand and gravel, the latter mainly subangular to subrounded flint, quartz and quartzite	0.9	(3.0)	1.0	(3.5)
	Clay, sandy, with fine to cobble-sized flint, quartz and quartzite	2.3	(7.5)	3.3	(11.0)
Upper Chalk	Chalk	0.5+	(1.5+)	3.8	(12.5)

SU 78 NE 7 | 7597 8842 | Near Upper Woodend Farm, Hambleden

Surface level (+181.7 m) +596 ft
Water not struck
Shell and auger (modified) 6 in (152 mm) diameter
February 1973

Waste 1.9 m (6.0 ft)
Bedrock 0.5 m+ (1.5 ft+)

LOG

Geological Classification	Lithology	Thickness m	ft	Depth m	ft
	Soil	0.3	(1.0)	0.3	(1.0)
Clay-with-flints	Clay, silty, brownish yellow with coarse to cobble-sized flints	1.6	(5.0)	1.9	(6.0)
Upper Chalk	Chalk, with flints	0.5+	(1.5+)	2.4	(8.0)

SU 78 NE 8 | 7572 8711 | Sunnyclose, Fawley

Surface level (+151.8 m) +498 ft
Water not struck
Shell and auger (modified) 6 in (152 mm) diameter
August 1972

Waste 2.2 m (7.0 ft)
Bedrock 0.5 m+ (1.5 ft+)

LOG

Geological Classification	Lithology	Thickness m	ft	Depth m	ft
	Soil	0.2	(0.5)	0.2	(0.5)
Clay-with-flints	Clay, sandy and pebbly, light brown	0.2	(0.5)	0.4	(1.5)
	Clay, silty, dark brown and orange, with flints	1.4	(4.5)	1.8	(6.0)
	Clay, dark brown, with flints and chalk pebbles	0.4	(1.5)	2.2	(7.0)
Upper Chalk	Chalk	0.5+	(1.5+)	2.7	(9.0)

SU 78 NE 9 7768 8952 Flint Hall, Hambleden Block C

Surface level (+ 64.9 m) + 213 ft
Water struck at + 62.1 m
Shell and auger (modified) 6 in (152 mm) diameter
February 1973

Overburden 0.2 m (0.5 ft)
Mineral 4.1 m (13.5 ft)
Bedrock 0.9 m + (3.0 ft +)

LOG

Geological Classification	Lithology	Thickness m	Thickness ft	Depth m	Depth ft
	Soil	0.2	(0.5)	0.2	(0.5)
Younger River Gravels (undifferentiated)	Gravel Gravel: mainly coarse with some fine, subangular to well rounded flint with some chalk and quartz below 2.3 m Sand: coarse and medium, pale brown, becoming clayey with depth	4.1	(13.5)	4.3	(14.0)
Middle Chalk	Chalk, soft and clayey in top 0.7 m, becoming flinty and hard with depth	0.9 +	(3.0 +)	5.2	(17.0)

GRADING

Mean for deposit

	%	mm	%
Gravel	72	+16	47
		−16+4	25
Sand	19	−4+1	10
		$-1+\frac{1}{4}$	8
		$-\frac{1}{4}+1/16$	1
Fines	9	−1/16	9

Bulk samples

Depth below surface (m)	Percentages Fines	Sand	Gravel
0.2 − 1.1	8	16	76
1.1 − 2.3	4	13	83
2.3 − 3.3	9	17	74
3.3 − 4.3	16	30	54

SU 78 NE 10 | 7740 8551 | Greenlands, near Mill End | Block C

Surface level (+ 36.5 m) + 120 ft
Water struck at + 30.3 m
Shell and auger (modified) 6 in (152 mm) diameter
August 1972

Overburden 2.3 m (7.5 ft)
Mineral 3.0 m (10.0 ft)
Waste 9.3 m (30.5 ft)
Bedrock 0.5 m+ (1.5 ft+)

LOG

Geological Classification	Lithology	Thickness m	ft	Depth m	ft
	Topsoil and made ground	0.4	(1.5)	0.4	(1.5)
Younger Coombe Deposits	Clay, silty, dark brown, with flint nodules	1.9	(6.0)	2.3	(7.5)
Younger River Gravels (undifferentiated)	'Very clayey' gravel Gravel: coarse with fine, subangular to subrounded flint with rounded quartz and chalk and with some limestone and ironstone Sand: coarse and medium with a little fine, angular to subangular flint and subrounded to rounded quartz; light brown to yellowish brown	3.0	(10.0)	5.3	(17.5)
	Silt, very gravelly, with sand	3.2	(10.5)	8.5	(28.0)
	Clay, silty, with a little fine sand in upper part, dark greyish brown to brown, ferruginous staining in parts	4.3	(14.0)	12.8	(42.0)
	Gravel, sandy, light brown, coarse, mainly flint with a little chalk	1.8	(6.0)	14.6	(28.0)
Middle Chalk	Chalk	0.5+	(1.5+)	15.1	(49.5)

GRADING

Mean for deposit

	%	mm	%
Gravel	58	+16	34
		−16+4	24
Sand	18	−4+1	8
		$-1+\frac{1}{4}$	18
		$-\frac{1}{4}+1/16$	2
Fines	24	−1/16	24

Bulk samples

Depth below surface (m)	Percentages Fines	Sand	Gravel
2.3 – 3.3	22	16	62
3.3 – 4.3	20	17	63
4.3 – 5.3	31	20	49

COMPOSITION

Depth below surface (m)	Percentages (in 4.75 to 9.5 mm fractions) Flint	Quartz	Ironstone	Chalk	Limestone	Miscellaneous
2.3 – 3.3	82	9	0	9	0	0
3.3 – 4.3	89	4	0	7	0	0
4.3 – 5.3	84	3	1	10	0	2
Mean	85	5	0	9	0	1

Surface level (+ 50.0 m) + 164 ft
Water struck at + 48.4 m
Shell and auger (modified) 6 in (152 mm) diameter
January 1973

Overburden 0.2 m (0.5 ft)
Mineral 6.5 m (21.5 ft)
Bedrock 0.6 m+ (2.0 ft+)

LOG

Geological Classification	Lithology	Thickness m	Thickness ft	Depth m	Depth ft
	Soil	0.2	(0.5)	0.2	(0.5)
Younger River Gravels (undifferentiated)	Gravel Gravel: coarse with fine and with cobbles, angular to subangular flint with some subangular to well rounded quartz Sand: mainly coarse with some medium and a little fine, pale grey	6.5	(21.5)	6.7	(22.0)
Middle Chalk	Chalk, soft in top 0.4 m, becoming hard with depth	0.6+	(2.0+)	7.3	(24.0)

GRADING

Mean for deposit

	%	mm	%
Gravel	74	+16	41
		−16+4	33
Sand	21	−4+1	12
		−1+¼	7
		−¼+1/16	2
Fines	5	−1/16	5

Bulk samples

Depth below surface (m)	Percentages Fines	Sand	Gravel
0.2 − 1.2	6	23	71
*1.2 − 2.2	5	25	70
*2.2 − 3.2	6	10	84
*3.2 − 4.2	3	25	72
*4.2 − 5.2	4	16	80
*5.2 − 6.2	6	21	73
*6.2 − 6.7	11	31	58

SU 78 NE 12 | 7858 8509 | Yewden Manor, Mill End | Block C

Surface level (+ 31.4 m) + 103 ft
Water struck at + 29.8 m
Shell and auger (modified) 6 in (152 mm) diameter
January 1973

Overburden 0.4 m (1.5 ft)
Mineral 7.1 m (23.5 ft)
Bedrock 0.5 m + (1.5 ft +)

LOG

Geological Classification	Lithology	Thickness m	Thickness ft	Depth m	Depth ft
	Soil, with flints	0.4	(1.5)	0.4	(1.5)
Younger River Gravels (undifferentiated)	Gravel Gravel: fine to coarse with cobbles, angular to well rounded, flint, with quartz, quartzite, chalk and some limestone Sand: medium and coarse, angular to subrounded flint and quartz, pale grey	7.1	(23.5)	7.5	(24.5)
Middle Chalk	Chalk	0.5+	(1.5+)	8.0	(26.0)

GRADING

Mean for deposit

	%	mm	%
Gravel	64	+16	33
		−16+4	31
Sand	29	−4+1	14
		$-1+\frac{1}{4}$	13
		$-\frac{1}{4}+1/16$	2
Fines	7	−1/16	7

Bulk samples

Depth below surface (m)	Percentages Fines	Sand	Gravel
0.4 − 1.3	14	14	72
*1.3 − 2.3	19	33	48
*2.3 − 3.3	3	31	66
*3.3 − 4.3	1	29	70
*4.3 − 5.3	4	30	66
*5.3 − 6.3	4	32	64
*6.3 − 7.5	5	36	59

SU 78 NE 13   7934 8894   Pheasants, Hambleden

Surface level (+161.5 m) + 530 ft
Water not struck
Shell and auger (modified) 6 in (152 mm) diameter
February 1973

Waste 1.7 m (5.5 ft)
Bedrock 0.5 m+ (1.5 ft+)

LOG

Geological Classification	Lithology	Thickness m	ft	Depth m	ft
	Soil	0.1	(0.5)	0.1	(0.5)
Sand and Gravel of unknown age	Clay, sandy, brown, and fine to cobble-sized, subangular to subrounded flint	1.6	(5.0)	1.7	(5.5)
Upper Chalk	Chalk, with flints	0.5+	(1.5+)	2.2	(7.0)

SU 78 NE 14   7977 8761   South of Rockwell End, Hambleden

Surface level (+139 m) +456 ft
Water not struck
Shell and auger (modified) 6 in (152 mm) diameter
August 1972

Waste 4.9 m (16.0 ft)
Bedrock 0.5 m+ (1.5 ft+)

LOG

Geological Classification	Lithology	Thickness m	ft	Depth m	ft
	Soil	0.3	(1.0)	0.3	(1.0)
	Subsoil, gravelly	0.8	(2.5)	1.1	(3.5)
Clay-with-flints	Clay, dark reddish brown, with flints	3.8	(12.5)	4.9	(16.0)
Upper Chalk	Chalk	0.5+	(1.5+)	5.4	(17.5)

SU 78 NE 15 | 7966 8600 | Near Burrow Farm, Hambleden

Surface level (+110.9 m) +364 ft
Water not struck
Shell and auger (modified) 6 in (152 mm) diameter
August 1972

Waste 8.0 (26.0 ft)
Bedrock 0.5 m+ (1.5 ft+)

LOG

Geological Classification	Lithology	Thickness m	Thickness ft	Depth m	Depth ft
	Soil	0.3	(1.0)	0.3	(1.0)
	Subsoil, gravelly	0.6	(2.0)	0.9	(3.0)
Older River Gravels (undifferentiated)	Clay, silty, reddish brown, mottled grey and orange-brown with pebbles of flint, quartz, red sandstone and green siltstone	5.6	(18.0)	6.5	(21.5)
	Sand, reddish brown, with a trace of flint gravel	1.3	(4.5)	7.8	(25.5)
	Clay, silty, dark reddish brown, with flints	0.2	(0.5)	8.0	(26.0)
Upper Chalk	Chalk	0.5+	(1.5+)	8.5	(28.0)

SU 78 SW 4 | 7030 8419 | Highmoor Cross, Highmoor

Surface level +136.2 m (+447 ft)
Water not struck
Shell and auger (modified) 6 in (152 mm) diameter
July 1972

Waste 3.1 m (10.0 ft)
Bedrock 1.0 m+ (3.5 ft+)

LOG

Geological Classification	Lithology	Thickness m	Thickness ft	Depth m	Depth ft
	Soil	0.2	(0.5)	0.2	(0.5)
Sand in Clay-with-flints	Sand, yellowish grey to orange-brown with scattered rounded flint and quartzite pebbles	0.4	(1.5)	0.6	(2.0)
Clay-with-flints	Clay, sandy and silty, reddish and orange-brown and fine to medium pebbles of flint, quartz and quartzite	2.5	(8.0)	3.1	(10.0)
? Reading Beds (not shown on resource map).	Sand, medium grained, bright reddish orange	1.0+	(3.5+)	4.1	(13.5)

SU 78 SW 5 | 7042 8156 | Peppard Common, Rotherfield Peppard | Block A

Surface level +163.3 m (+339 ft)
Water not struck
Shell and auger (modified) 6 in (152 mm) diameter
July 1972

Overburden 0.1 m (0.5 ft)
Mineral 8.8 m (29.0 ft)
Waste 0.3 m (1.0 ft)
Bedrock 0.5 m+ (1.5 ft+)

LOG

Geological Classification	Lithology	Thickness m	Thickness ft	Depth m	Depth ft
	Soil	0.1	(0.5)	0.1	(0.5)
Older River Gravels (undifferentiated)	'Very clayey' gravel Gravel: coarse with fine with some cobbles, subrounded to rounded flint with some rounded quartz Sand: medium and coarse with some fine, fairly clayey throughout, flint and quartz, orange brown	8.8	(29.0)	8.9	(29.0)
	Clay, silty with occasional pebbles, brown	0.3	(1.0)	9.2	(30.0)
Upper Chalk	Chalk, white, dirty, containing harder chalk nodules	0.5+	(1.5+)	9.7	(32.0)

GRADING

Mean for deposit

	%	mm	%
Gravel	53	+16	30
		−16+4	23
Sand	20	−4+1	8
		−1+¼	9
		−¼+1/16	3
Fines	27	−1/16	27

Bulk samples

Depth below surface (m)	Percentages Fines	Sand	Gravel
0.1 − 0.8	11	8	81
0.8 − 1.8	28	19	53
1.8 − 2.8	28	30	42
2.8 − 3.8	24	21	55
3.8 − 4.8	29	21	50
4.8 − 5.9	31	23	46
5.9 − 6.8	46	10	44
6.8 − 7.6	23	23	55
7.6 − 8.9	22	18	60

SU 78 SW 6	7033 8058	Bishopswood Farm, Sonning Common	Block A

Surface level + 97.8 m (+ 321 ft)
Water not struck
Shell and auger (modified) 6 in (152 mm) diameter
August 1972

Overburden 0.3 m (1.0 ft)
Mineral 5.9 m (19.5 ft)
Bedrock 0.5 m + (1.5 ft +)

LOG

Geological Classification	Lithology	Thickness m	Thickness ft	Depth m	Depth ft
	Soil	0.3	(1.0)	0.3	(1.0)
Older River Gravels (undifferentiated)	'Very clayey' gravel Gravel: fine to coarse with occasional cobbles, subangular to rounded flint with rounded quartz and with some subrounded to rounded ? Greensand pebbles Sand: mainly medium with some coarse, subrounded to rounded quartz with some angular to subangular flint, orange brown	5.9	(19.5)	6.2	(20.5)
Upper Chalk	Chalk, soft, white, with harder nodules and with large flint nodules	0.5+	(1.5+)	6.7	(22.0)

GRADING

Mean for deposit

	%	mm	%
Gravel	49	+16	29
		−16+4	20
Sand	32	−4+1	9
		$-1+\frac{1}{4}$	21
		$-\frac{1}{4}+1/16$	2
Fines	19	−1/16	19

Bulk samples

Depth below surface (m)	Percentages Fines	Sand	Gravel
0.3 − 1.3	37	31	32
1.3 − 2.3	13	29	58
2.3 − 3.3	24	38	38
3.3 − 4.3	16	31	53
4.3 − 5.3	9	37	54
5.3 − 6.2	14	27	59

SU 78 SW 7 7164 8478 Bromsden Cottages, Highmoor Block A

Surface level +130.1 m (+427 ft)
Water not struck
Shell and auger (modified) 6 in (152 mm) diameter
July 1972

Overburden 0.3 m (1.0 ft)
Mineral 3.8 m (12.5 ft)
Waste 0.7 m (2.5 ft)
Bedrock 0.5 m+ (1.5 ft+)

LOG

Geological Classification	Lithology	Thickness m	ft	Depth m	ft
	Soil, sandy	0.3	(1.0)	0.3	(1.0)
Older River Gravels (undifferentiated)	'Clayey' gravel Gravel: mainly coarse with some fine and with scattered cobbles, subrounded to rounded quartzite and flint Sand: mainly medium with a little coarse and fine, quartz and flint	3.8	(12.5)	4.1	(13.5)
	Clay, silty, sandy with occasional irregular flints, reddish brown to light brown	0.7	(2.5)	4.8	(15.5)
Upper Chalk	Chalk, firm, light grey, streaked reddish brown	0.5+	(1.5+)	5.3	(17.5)

GRADING

Mean for deposit

	%	mm	%
Gravel	52	+16	34
		−16+4	18
Sand	28	−4+1	5
		−1+¼	19
		−¼+1/16	4
Fines	20	−1/16	20

Bulk samples

Depth below surface (m)	Percentages Fines	Sand	Gravel
0.3 − 1.3	20	23	57
1.3 − 2.3	20	36	44
2.3 − 3.3	23	22	55
3.3 − 4.1	14	32	54

SU 78 SW 8 7115 8308 Near Satwell Spinneys, Shepherd's Green

Surface level +128 m (+420 ft)
Water not struck
Shell and auger (modified) 6 in (152 mm) diameter
August 1972

Waste 2.4 m (8.0 ft)
Bedrock 0.5 m+ (1.5 ft+)

LOG

Geological Classification	Lithology	Thickness m	Thickness ft	Depth m	Depth ft
	Soil	0.3	(1.0)	0.3	(1.0)
Older River Gravels	Clay, pebbly, dark reddish brown	2.1	(7.0)	2.4	(8.0)
Upper Chalk	Chalk	0.5+	(1.5+)	2.9	(9.5)

SU 78 SW 9 7127 8234 Near Grasslands, Rotherfield Peppard

Surface level +103.6 m (+340 ft)
Water not struck
Shell and auger (modified) 6 in (152 mm) diameter
August 1972

Waste 5.3 m (17.5 ft)
Bedrock 0.5 m+ (1.5 ft+)

LOG

Geological Classification	Lithology	Thickness m	Thickness ft	Depth m	Depth ft
	Soil	0.2	(0.5)	0.2	(0.5)
Older River Gravels (undifferentiated)	Clay, silty, sandy with flints, light brown	0.8	(2.5)	1.0	(3.5)
	Clay with large nodular flints and occasional rounded vein quartz, dark brown	4.3	(14.0)	5.3	(17.5)
Upper Chalk	Chalk, soft, white, with thin brown clay seams in upper part and nodular flints	0.5+	(1.5+)	5.8	(19.0)

SU 78 SW 10 | 7190 8045 | Blountscourt Farm, Sonning Common | Block A

Surface level +93.9 m (+308 ft)
Water not struck
Shell and auger (modified) 6 in (152 mm) diameter
July 1972

Overburden 0.1 m (0.5 ft)
Mineral 2.4 m (8.0 ft)
Waste 4.5 m (15.0 ft)
Mineral 5.2 m (17.0 ft)
Bedrock 0.5 m+ (1.5 ft+)

LOG

Geological Classification		Lithology	Thickness m	Thickness ft	Depth m	Depth ft
		Soil	0.1	(0.5)	0.1	(0.5)
Older River Gravels (undifferentiated)	(a)	'Very clayey' gravel Gravel: mainly coarse with some fine, subrounded to rounded flint, with some rounded quartz and quartzite Sand: medium with some coarse, mainly subrounded to rounded quartz	2.4	(8.0)	2.5	(8.0)
		Clay, sandy, silty with pebbles orange-brown	4.5	(15.0)	7.0	(23.0)
	(b)	'Clayey gravel Gravel: mainly coarse with some fine flint, occasional cobbles near the base Sand: mainly medium with a little coarse	5.2	(17.0)	12.2	(40.0)
Upper Chalk		Chalk, with flint nodules	0.5+	(1.5+)	12.7	(41.5)

GRADING

Mean for deposit

	%	mm	%
Gravel	51	+16	33
		−16+4	18
Sand	26	−4+1	7
		$-1+\frac{1}{4}$	17
		$-\frac{1}{4}+\frac{1}{2}$	2
Fines	23	−1/16	23

Bulk samples

	Depth below surface (m)	Percentages Fines	Sand	Gravel
(a)	0.1 − 1.0	33	21	46
	1.0 − 1.8	41	20	39
	1.8 − 2.5	29	26	45
	Mean	35	22	43
(b)	7.0 − 8.0	28	29	43
	8.0 − 9.0	20	35	45
	9.0 − 10.0	18	34	48
	10.0 − 11.0	12	27	61
	11.0 − 12.2	9	16	75
	Mean	17	28	55

SU 78 SW 11 7279 8471 Lawrence's Farm Cottage, Bix

Surface level +129.8 m (+426 ft)
Water not struck
Shell and auger (modified) 6 in (152 mm) diameter
July 1972

Waste 0.2 m (0.5 ft)
Bedrock 1.8 m+ (6.0 ft+)

LOG

Geological Classification	Lithology	Thickness m	Thickness ft	Depth m	Depth ft
	Soil	0.2	(0.5)	0.2	(0.5)
Reading Beds	Clay, pebbly, light yellowish-grey and light brown with patches of orange-brown	0.8	(2.5)	1.0	(3.5)
	Clay, very sandy, varying to clayey sand, bright reddish-brown streaked pale greenish-grey	1.0+	(3.5+)	2.0	(6.5)

SU 78 SW 12 7249 8395 Brickfield Cottages, Rotherfield Greys

Surface level +133.8 m (439 ft)
Water not struck
Shell and auger (modified) 6 in (152 mm) diameter
August 1972

Waste 1.7 m (5.5 ft)
Bedrock 0.5 m+ (1.5 ft+)

LOG

Geological Classification	Lithology	Thickness m	Thickness ft	Depth m	Depth ft
	Soil	0.2	(0.5)	0.2	(0.5)
Clay-with-flints	Clay, silty with flints, light brown becoming dark reddish brown	1.5	(5.0)	1.7	(5.5)
Upper Chalk	Chalk	0.5+	(1.5+)	2.2	(7.0)

SU 78 SW 13 7219 8278 Greysgreen Farm, Greys Green Block A

Surface level +98.8 m (+324 ft)
Water not struck
Shell and auger (modified) 6 in (152 mm) diameter
July 1972

Overburden 1.8 (6.0 ft)
Mineral 2.1 m (7.0 ft)
Waste 7.5 m (24.5 ft)
Bedrock 0.5 m+ (1.5 ft+)

LOG

Geological Classification	Lithology	Thickness m	Thickness ft	Depth m	Depth ft
	Soil	0.3	(1.0)	0.3	(1.0)
Older River Gravels (undifferentiated)	Silt, sandy, gravelly, light brown	0.4	(1.5)	0.7	(2.5)
	Clay, silty with occasional flint and rounded vein quartz pebbles	1.1	(3.5)	1.8	(6.0)
	'Very clayey' gravel Gravel: mainly coarse, some fine and occasional cobbles, subangular to subrounded with some rounded flint, with a trace of quartzite Sand: mainly medium with some coarse and fine, mainly subrounded to rounded quartz	2.1	(7.0)	3.9	(13.0)
	Clay, silty, with nodular flints, red-brown, becoming dark yellowish brown	7.5	(24.5)	11.4	(37.5)
Upper Chalk	Chalk, soft, white with nodular flints and clay seams	0.5+	(1.5+)	11.9	(39.0)

GRADING

Mean for deposit

	%	mm	%
Gravel	55	+16	39
		−16+4	16
Sand	20	−4+1	5
		$-1+\frac{1}{4}$	12
		$-\frac{1}{4}+1/16$	3
Fines	25	−1/16	25

Bulk samples

Depth below surface (m)	Percentages Fines	Sand	Gravel
1.8 – 2.8	25	18	57
2.8 – 3.9	25	21	54

SU 78 SW 14	7265 8200	Near Black Cottages, Rotherfield Greys	Block A

Surface level +98.1 m (+322 ft)
Water not struck
Shell and auger (modified) 6 in (152 mm) diameter
August 1972

Overburden 4.0 m (13.0 ft)
Mineral 12.1 m (39.5 ft)
Waste 0.7 m (2.5 ft)
Bedrock 0.5 m+ (1.5 ft+)

LOG

Geological Classification	Lithology	Thickness m	Thickness ft	Depth m	Depth ft
	Soil	0.1	(0.5)	0.1	(0.5)
Older River Gravels (undifferentiated)	Clay, pebbly, silty, orange brown, becoming dark brown with depth	2.8	(9.0)	2.9	(9.5)
	Clay, sandy, silty, orange brown	1.1	(3.5)	4.0	(13.0)
	'Clayey' gravel, with clay band or lens between 5.1 m and 5.8 m Gravel: fine to coarse, subangular to subrounded with rounded flint, with some rounded vein quartz and a trace of rounded Greensand siltstones Sand: mainly medium with some coarse, and fine, subrounded to rounded quartz	12.1	(39.5)	16.1	(53.0)
	Clay, with large nodular flints, orange brown to dark brown	0.7	(2.5)	16.8	(55.0)
Upper Chalk	Chalk, white with clay seams in upper part	0.5+	(1.5+)	17.3	(57.0)

GRADING

Mean for deposit

	%	mm	%
Gravel	49	+16	27
		−16+4	22
Sand	37	−4+1	8
		$-1+\frac{1}{4}$	25
		$-\frac{1}{4}+1/16$	4
Fines	14	−1/16	14

Bulk samples

Depth below surface (m)	Percentages Fines	Sand	Gravel
4.0 − 5.1	19	31	50
5.8 − 6.8	18	33	49
6.8 − 7.8	21	29	50
7.8 − 8.8	16	36	48
8.8 − 9.8	16	43	41
9.8 − 10.8	17	39	44
10.8 − 11.8	13	41	46
11.8 − 12.8	8	39	53
12.8 − 13.8	11	33	56
13.8 − 14.8	18	39	43
14.8 − 16.1	9	34	57

SU 78 SW 15 7325 8326 Near Farm, Rotherfield Greys

Surface level +105.2 m (+345 ft)
Water not struck
Shell and auger (modified) 6 in (152 mm) diameter
November 1972

Waste 13.7 (45.0 ft)
Bedrock 0.5 m+ (1.5 ft+)

LOG

Geological Classification	Lithology	Thickness m	ft	Depth m	ft
	Soil	0.2	(0.5)	0.2	(0.5)
Older River Gravels (undifferentiated)	Clay, sandy, gravelly with some cobbles, brown to reddish brown	13.5	(44.5)	13.7	(45.0)
Upper Chalk	Chalk, with flints	0.5+	(1.5+)	14.2	(46.5)

SU 78 SW 16 7339 8205 Cowsfields Farm, Rotherfield Greys Block A

Surface level +93.9 m (+308 ft)
Water not struck
Shell and auger (modified) 6 in (152 mm) diameter
August 1972

Overburden 0.2 m (0.5 ft)
Mineral 3.0 m (10.0 ft)
Bedrock 0.5 m+ (1.5 ft+)

LOG

Geological Classification	Lithology	Thickness m	ft	Depth m	ft
	Soil	0.2	(0.5)	0.2	(0.5)
Older River Gravels (undifferentiated)	'Very clayey' gravel Gravel: fine and coarse, coarsening downwards, subangular to subrounded, with rounded flint, with rounded vein quartz and with Greensand siltstones Sand: mainly medium with some coarse, mainly subrounded to rounded quartz, orange brown	3.0	(10.0)	3.2	(10.5)
Upper Chalk	Chalk, white, putty-like	0.5+	(1.5+)	3.7	(12.0)

GRADING

Mean for deposit

	%	mm	%
Gravel	52	+16	27
		−16+4	25
Sand	27	−4+1	6
		$-1+\frac{1}{4}$	19
		$-\frac{1}{4}+1/16$	2
Fines	21	−1/16	21

Bulk samples

Depth below surface (m)	Percentages Fines	Sand	Gravel
0.2 − 0.7	37	28	35
0.7 − 1.7	30	24	46
1.7 − 2.7	7	34	59
2.7 − 3.2	14	21	65

SU 78 SW 17 7450 8307 Badgemoor House, Rotherfield Greys

Surface level + 93.8 m (+ 307 ft)
Water not struck
Shell and auger (modified) 6 in (152 mm) diameter
November 1972

Waste 5.1 m (17.0 ft)
Bedrock 0.5 m+ (1.5 ft+)

LOG

Geological Classification	Lithology	Thickness m	Thickness ft	Depth m	Depth ft
	Made ground	0.6	(2.0)	0.6	(2.0)
Older River Gravels (undifferentiated)	Gravel, sandy, clayey	0.7	(2.5)	1.3	(4.0)
	Clay, sandy, gravelly, gravel fine and coarse with some cobbles, brown	3.8	(12.5)	5.1	(17.0)
Upper Chalk	Chalk, with flints	0.5+	(1.5+)	5.6	(18.5)

SU 78 SW 18 7451 8150 Highlands Farm, Harpsden Block B

Surface level + 77.1 m (+ 253 ft)
Water struck at surface
Shell and auger (modified) 6 in (152 mm) diameter
November 1972

Overburden 0.2 m (0.5 ft)
Mineral 6.5 m (21.5 ft)
Bedrock 0.5 m+ (1.5 ft)

LOG

Geological Classification	Lithology	Thickness m	Thickness ft	Depth m	Depth ft
	Soil	0.2	(0.5)	0.2	(0.5)
Older River Gravels (undifferentiated)	'Clayey' gravel Gravel: mainly coarse with some fine, occasional cobbles, subangular to well rounded flint with quartz and quartzite Sand: mainly medium and coarse, brown, clayey throughout	6.5	(21.5)	6.7	(22.0)
Upper Chalk	Chalk with flints	0.5+	(1.5+)	7.2	(23.5)

GRADING

Mean for deposit

	%	mm	%
Gravel	59	+16	36
		−16+4	23
Sand	22	−4+1	9
		−1+¼	12
		−¼+1/16	1
Fines	19	−1/16	19

Bulk samples

Depth below surface (m)	Fines	Sand	Gravel
* 0.2 − 1.1	26	25	49
* 1.1 − 2.1	17	28	55
* 2.1 − 3.1	4	22	74
* 3.1 − 4.1	19	23	58
* 4.1 − 5.1	23	17	60
* 5.1 − 6.1	21	20	59
* 6.1 − 6.7	26	14	60

SU 78 SW 19	7472 8039	Perseverance Cottages, Harpsden	Block B

Surface level c+79.2 m (+260 ft)
Groundwater level not recorded
Shell and auger (modified) 6 in (152 mm) diameter
September 1974

Overburden 0.4 m (1.5 ft)
Mineral 2.4 m (8.0 ft)
Bedrock 0.5 m+ (1.5 ft+)

LOG

Geological Classification	Lithology	Thickness m	Thickness ft	Depth m	Depth ft
	Soil	0.3	(1.0)	0.3	(1.0)
Older River Gravels (undifferentiated)	Sand, very clayey, with flints, light brown	0.1	(0.5)	0.4	(1.5)
	'Very clayey' gravel Gravel: coarse with fine, angular to subrounded flint, with subrounded to rounded sandstone Sand: medium with some coarse and fine, clayey, reddish brown	2.4	(8.0)	2.8	(9.0)
Upper Chalk	Chalk, with flints	0.5+	(1.5+)	3.3	(11.0)

GRADING

Mean for deposit

	%	mm	%
Gravel	49	+16	32
		−16+4	17
Sand	27	−4+1	7
		−1+$\frac{1}{4}$	12
		−$\frac{1}{4}$+1/16	8
Fines	24	−1/16	24

Bulk samples

Depth below surface (m)	Percentages Fines	Sand	Gravel
0.4 − 1.4	35	30	35
1.4 − 2.8	16	24	60

SU 78 SE 14 | 7550 8432 | No Man's Hill, North of Henley | Block A

Surface level +97.5 m (+320 ft)
Water not struck
Shell and auger (modified) 6 in (152 mm) diameter
August 1972

Overburden 0.2 (0.5 ft)
Mineral 9.5 m (31.0 ft)
Bedrock 0.5 m+ (1.5 ft+)

LOG

Geological Classification	Lithology	Thickness m	Thickness ft	Depth m	Depth ft
	Soil	0.2	(0.5)	0.2	(0.5)
Older River Gravels (undifferentiated)	'Clayey' gravel Gravel: fine to coarse with occasional cobbles, subangular to subrounded with rounded flint with some rounded vein quartz Sand: medium and coarse, subrounded to rounded quartz with some angular to subangular flint	9.5	(31.0)	9.7	(32.0)
Upper Chalk	Chalk with flints	0.5+	(1.5+)	10.2	(33.5)

GRADING

Mean for deposit

	%	mm	%
Gravel	62	+16	35
		−16+4	25
Sand	21	−4+1	8
		−1+$\frac{1}{4}$	12
		−$\frac{1}{4}$+1/16	1
Fines	17	−1/16	17

Bulk samples

Depth below surface (m)	Percentages Fines	Sand	Gravel
0.2 − 0.9	18	18	64
0.9 − 1.8	13	26	61
1.8 − 2.8	24	14	62
2.8 − 3.8	6	13	81
3.8 − 4.8	8	17	75
4.8 − 5.8	7	19	74
5.8 − 6.8	15	21	64
6.8 − 7.8	25	26	49
7.8 − 8.8	36	30	34
8.8 − 9.7	15	31	54

SU 78 SE 15 | 7537 8161 | Near Wootten Manor, Henley | Block B

Surface level +77.1 m (+253 ft)
Water not struck
Shell and auger (modified) 6 in (152 mm) diameter
August 1972

Overburden 2.8 m (9.0 ft)
Mineral 4.1 m (13.5 ft)
Waste 0.9 m (3.0 ft)
Bedrock 0.5 m+ (1.5 ft+)

LOG

Geological Classification	Lithology	Thickness m	Thickness ft	Depth m	Depth ft
	Soil	0.2	(0.5)	0.2	(0.5)
Older River Gravels (undifferentiated)	Clay, silty and sandy, orange brown, with scattered flint pebbles	2.6	(8.5)	2.8	(9.0)
	'Clayey' gravel Gravel: fine with some coarse, subangular to subrounded flint with some subrounded vein quartz Sand: mainly medium with some coarse and a little fine subrounded to rounded quartz with angular to subangular flint	4.1	(13.5)	6.9	(22.5)
	Clay, very sandy with a trace of gravel, dark brown with large nodular flint cobbles, near base	0.9	(3.0)	7.8	(25.5)
Upper Chalk	Chalk, clayey in upper part, with large nodular flint cobbles	0.5+	(1.5+)	8.3	(27.0)

GRADING

Mean for deposit

	%	mm	%
Gravel	51	+16	21
		−16+4	30
Sand	33	−4+1	9
		$-1+\frac{1}{4}$	21
		$-\frac{1}{4}+1/16$	3
Fines	16	−1/16	16

Bulk samples

Depth below surface (m)	Percentages Fines	Sand	Gravel
2.8 − 3.8	12	26	62
3.8 − 4.8	16	36	48
4.8 − 5.8	15	34	51
5.8 − 6.9	22	33	45

SU 78 SE 16 7565 8017 Near Red Hatch, Harpsden

Surface level +81.6 m (+268 ft)
Water not struck
Shell and auger (modified) 6 in (152 mm) diameter
November 1972

Waste 6.0 m (19.5 ft)
Bedrock 0.5 m+ (1.5 ft+)

LOG

Geological Classification	Lithology	Thickness m	Thickness ft	Depth m	Depth ft
	Soil	0.2	(0.5)	0.2	(0.5)
Clay-with-flints	Clay, sandy, with medium to cobble-sized flint, and some quartz and quartzite pebbles. Chalk pebbles towards base	5.8	(19.0)	6.0	(19.5)
Upper Chalk	Chalk, with large flints	0.5+	(1.5+)	6.5	(21.5)

SU 78 SE 17 | 7633 8480 | Near Fawley Court, North of Henley | Block C

Surface level +34.1 m (+112 ft)
Water struck at +29.3 m
Shell and auger (modified) 6 in (152 mm) diameter
August 1972

Overburden 0.2 m (0.5 ft)
Mineral 10.1 m (33.0 ft)
Waste 0.4 m (1.5 ft)
Bedrock 0.5 m+ (1.5 ft+)

LOG

Geological Classification	Lithology	Thickness m	Thickness ft	Depth m	Depth ft
	Soil	0.2	(0.5)	0.2	(0.5)
Younger Coombe Deposits passing into Younger River Gravels (undifferentiated)	'Clayey' gravel Gravel: mainly coarse, with some fine, subangular to subrounded with rounded flint, with some subrounded to rounded chalk in top 3 m and a trace of rounded quartz Sand: medium and coarse subrounded to rounded quartz with some subrounded to rounded chalk and subangular to angular flint	10.1	(33.0)	10.3	(34.0)
	Clay, sandy, with scattered pebbles	0.4	(1.5)	10.7	(35.0)
Middle Chalk	Chalk	0.5	(1.5+)	11.2	(36.5)

GRADING

Mean for deposit

	%	mm	%
Gravel	63	+16	39
		−16+4	24
Sand	20	−4+1	9
		$-1+\frac{1}{4}$	10
		$-\frac{1}{4}+1/16$	1
Fines	17	−1/16	17

Bulk samples

Depth below surface (m)	Percentages Fines	Sand	Gravel
0.2 – 1.2	17	18	65
1.2 – 2.2	17	17	66
2.2 – 3.0	20	18	62
3.0 – 4.0	28	17	55
*4.0 – 5.0	12	15	73
*5.0 – 6.0	14	18	68
*6.0 – 7.0	6	22	72
*7.0 – 8.0	21	13	66
*8.0 – 9.0	18	33	49
*9.0 – 10.3	19	30	51

SU 78 SE 18	7684 8430	Near the Rectory, Remenham	Block C

Surface level +31.4 m (+103 ft)
Water struck at +27.8 m
Shell and auger (modified) 6 in (152 mm) diameter
August 1972

Overburden 3.6 m (12.0 ft)
Mineral 5.7 m (18.5 ft)
Bedrock 0.5 m+ (1.5 ft+)

LOG

Geological Classification	Lithology	Thickness m	ft	Depth m	ft
	Made ground	1.3	(4.5)	1.3	(4.5)
Alluvium	Clay, silty, bluish green	2.3	(7.5)	3.6	(12.0)
Younger River Gravels (undifferentiated)	Gravel, Gravel: mainly coarse, with some fine and with occasional cobbles, becoming less coarse with depth, subangular to subrounded with rounded flint, with some rounded quartz and vein quartz, and with some chalk below 6.6 m Sand: coarse and medium, subrounded to rounded quartz with angular to subangular flint, grey brown in upper part changing with depth to brown	5.7	(18.5)	9.3	(30.5)
Middle Chalk	Chalk, white	0.5+	(1.5+)	9.8	(32.0)

GRADING

Mean for deposit

	%	mm	%
Gravel	69	+16	44
		−16+4	25
Sand	29	−4+1	16
		$-1+\frac{1}{4}$	12
		$-\frac{1}{4}+1/16$	1
Fines	2	−1/16	2

Bulk samples

Depth below surface (m)	Percentages Fines	Sand	Gravel
*3.6 − 4.6	1	13	86
*4.6 − 5.6	1	27	72
*5.6 − 6.6	2	31	67
*6.6 − 7.6	1	35	64
*7.6 − 8.6	2	35	63
*8.6 − 9.3	3	35	62

SU 78 SE 19   7623 8356   Near Swiss Farm, North of Henley   Block C

Surface level + 32.3 m (+ 106 ft)
Water struck at + 30.8 m
Shell and auger (modified) 6 in (152 mm) diameter
August 1972

Overburden 1.4 m (4.5 ft)
Mineral 10.8 m (35.5 ft)
Bedrock 0.5 m + (1.5 ft +)

LOG

Geological Classification	Lithology	Thickness m	Thickness ft	Depth m	Depth ft
	Soil	0.3	(1.0)	0.3	(1.0)
Alluvium	Clay, silty and sandy, dark brown, with scattered pebbles	1.1	(3.5)	1.4	(4.5)
Younger River Gravels (undifferentiated)	Gravel Gravel: fine to coarse, becoming less coarse with depth, subangular to rounded flint with rounded quartz and chalk and with some ironstone and limestone Sand: medium and coarse, subrounded to rounded quartz with angular to subangular flint	10.8	(35.5)	12.2	(40.0)
Middle Chalk	Chalk	0.5+	(1.5+)	12.7	(41.5)

GRADING

Mean for deposit

	%	mm	%
Gravel	62	+16	34
		−16+4	28
Sand	28	−4+1	11
		$-1+\frac{1}{4}$	15
		$-\frac{1}{4}+1/16$	2
Fines	10	−1/16	10

Bulk samples

Depth below surface (m)	Fines	Sand	Gravel
*1.4 − 2.4	14	18	68
*2.4 − 3.4	4	22	74
*3.4 − 4.4	2	9	89
*4.4 − 5.4	10	29	61
*5.4 − 6.4	7	27	66
*6.4 − 7.4	10	27	63
*7.4 − 8.4	13	40	47
*8.4 − 9.4	12	32	56
*9.4 − 10.4	11	34	55
*10.4 − 11.4	14	36	50
*11.4 − 12.2	12	39	49

COMPOSITION

Percentages (in 4.75 to 9.5 mm fractions)

Depth below surface (m)	Flint	Quartz	Ironstone	Chalk	Limestone
1.4 − 2.4	91	5	1	2	1
2.4 − 3.4	87	6	1	1	5
3.4 − 4.4	83	6	1	8	2
4.4 − 5.4	86	2	1	6	5
5.4 − 6.4	82	8	3	5	2
6.4 − 7.4	76	4	2	15	3
7.4 − 8.4	82	4	1	12	1
8.4 − 9.4	84	6	1	8	1
9.4 − 10.4	77	5	2	15	1
10.4 − 11.4	90	2	1	6	1
11.4 − 12.2	79	8	4	8	1
Mean	83	5	2	8	2

SU 78 SE 20 7679 8261 Near Matson House, East of Henley Block C

Surface level +32.9 m (+108 ft)
Water struck at +30.0 m
Shell and auger (modified) 6 in (152 mm) diameter
August 1972

Overburden 2.9 m (9.5 ft)
Mineral 3.0 m (10.0 ft)
Waste 0.3 m (1.0 ft)
Bedrock 0.5 m+(1.5 ft+)

LOG

Geological Classification	Lithology	Thickness m	Thickness ft	Depth m	Depth ft
	Made ground	1.0	(3.5)	1.0	(3.5)
Younger River Gravels (undifferentiated)	Silt with some sand and gravel	1.9	(6.0)	2.9	(9.5)
	'Clayey' gravel Gravel: fine to coarse with occasional cobbles, subangular to subrounded, occasionally rounded flint, with rounded hard chalk pebbles below 5.5 m, and with some rounded vein quartz Sand: medium and coarse, subrounded to rounded quartz with some angular to subangular flint	3.0	(10.0)	5.9	(19.5)
	Sandy, chalk gravel	0.3	(1.0)	6.2	(20.5)
Upper Chalk	Chalk	0.5+	(1.5+)	6.7	(22.0)

GRADING

Mean for deposit

	%	mm	%
Gravel	60	+16	32
		−16+4	28
Sand	28	−4+1	11
		$-1+\frac{1}{4}$	15
		$-\frac{1}{4}+1/16$	2
Fines	12	−1/16	12

Bulk samples

Depth below surface (m)	Percentages Fines	Sand	Gravel
*2.9 − 3.9	27	31	42
*3.9 − 4.9	2	16	82
*4.9 − 5.9	8	37	55

SU 78 SE 21 — 7749 8456 — Near Remenham — Block C

Surface level +33.2 m (+109 ft)
Water struck at +31.2 m
Shell and auger (modified) 6 in (152 mm) diameter
January 1973

Overburden 0.9 m (3.0 ft)
Mineral 8.4 m (27.5 ft)
Bedrock 0.5 m+ (1.5 ft+)

LOG

Geological Classification	Lithology	Thickness m	Thickness ft	Depth m	Depth ft
	Soil	0.2	(0.5)	0.2	(0.5)
Younger River Gravels (undifferentiated)	Clay, sandy with some pebbles of flint, dark brown	0.7	(2.5)	0.9	(3.0)
	'Clayey' gravel Gravel: fine to coarse, subangular to subrounded, black and brown flint with mainly subrounded quartz and quartzite and well rounded chalk Sand: medium with coarse, clayey, to 4.0 m depth, light brown to yellowish brown	8.4	(27.5)	9.3	(30.5)
Middle Chalk	Chalk	0.5+	(1.5+)	9.8	(32.0)

GRADING

Mean for deposit

	%	mm	%
Gravel	47	+16	23
		−16+4	24
Sand	40	−4+1	13
		$-1+\frac{1}{4}$	25
		$-\frac{1}{4}+1/16$	2
Fines	13	−1/16	13

Bulk samples

Depth below surface (m)	Percentages Fines	Sand	Gravel
0.9 − 2.0	20	40	40
*2.0 − 3.0	28	70	2
*3.0 − 4.0	26	33	41
*4.0 − 5.0	8	34	58
*5.0 − 6.0	9	26	65
*6.0 − 7.0	6	32	62
*7.0 − 8.0	5	46	49
*8.0 − 9.3	2	36	62

SU 78 SE 22 | 7778 8317 | Near Common Barn, Remenham Hill | Block A

Surface level +98.1 m (+322 ft)
Water not struck
Shell and auger (modified) 6 in (152 mm) diameter
August 1972

Overburden 0.2 m (0.5 ft)
Mineral 3.0 m (10.0 ft)
Waste 2.4 m (8.0 ft)
Mineral 4.7 m (15.5 ft)
Waste 0.9 m (3.0 ft)
Bedrock 0.5 m+ (1.5 ft+)

LOG

Geological Classification		Lithology	Thickness m	Thickness ft	Depth m	Depth ft
		Soil	0.2	(0.5)	0.2	(0.5)
Older River Gravels (undifferentiated)	(a)	'Very clayey' gravel Gravel: coarse with fine, with occasional cobbles, subangular to rounded flint with rounded vein quartz Sand: mainly medium, with a little coarse and fine, subrounded to rounded quartz with angular to sub-angular flint	3.0	(10.0)	3.2	(10.5)
		Silt, sandy, gravelly, reddish brown	2.4	(8.0)	5.6	(18.5)
	(b)	'Very clayey' gravel Gravel: coarse with fine, with occasional cobbles, subrounded to rounded flint, with some rounded sandstone and rounded vein quartz Sand: mainly medium, with a little coarse and fine, subrounded to rounded quartz with angular to sub-angular flint, orange to reddish brown	4.7	(15.5)	10.3	(34.0)
		Clay, silty, with flints, very dark brown, with manganese staining	0.9	(3.0)	11.2	(36.5)
Upper Chalk		Chalk	0.5+	(1.5+)	11.7	(38.5)

GRADING

Mean for deposit

	%	mm	%
Gravel	52	+16	32
		−16+4	26
Sand	25	−4+1	5
		$-1+\frac{1}{4}$	18
		$-\frac{1}{4}+1/16$	2
Fines	23	−1/16	23

Bulk samples

	Depth below surface (m)	Percentages Fines	Sand	Gravel
(a)	0.2 − 1.4	14	17	69
	1.4 − 2.4	22	16	62
	2.4 − 3.2	34	18	48
	Mean	22	17	61
(b)	5.6 − 6.6	34	37	29
	6.6 − 7.6	28	29	43
	7.6 − 8.6	28	36	36
	8.6 − 9.6	10	29	61
	9.6 − 10.3	11	20	69
	Mean	23	31	46

SU 78 SE 23 7728 8148 Near Sheephouse Farm, Henley Block C

Surface level (+ 32.6 m) + 107 ft
Water struck at + 29.8 m
Shell and auger (modified) 6 in (152 mm) diameter
August 1972

Overburden 2.2 m (7.0 ft)
Mineral 9.1 m (30.0 ft)
Bedrock 0.5 m+ (1.5 ft+)

LOG

Geological Classification	Lithology	Thickness m	Thickness ft	Depth m	Depth ft
	Soil	0.3	(1.0)	0.3	(1.0)
Alluvium	Clay, sandy, orange-brown, with scattered small chalk pebbles	0.9	(3.0)	1.2	(4.0)
	Silt, with sand, orange-brown	1.0	(3.5)	2.2	(7.0)
Younger River Gravels (undifferentiated)	'Clayey' sandy gravel Gravel: fine to coarse, subangular to subrounded with rounded flint, with quartz, chalk and limestone and with some ironstone Sand: medium with some coarse and and fine, subrounded to rounded quartz, with subangular to subrounded flint, orange brown	9.1	(30.0)	11.3	(37.0)
Upper Chalk	Chalk	0.5+	(1.5+)	11.8	(38.5)

GRADING

Mean for deposit

	%	mm	%
Gravel	37	+16	20
		−16+4	17
Sand	45	−4+1	7
		$-1+\frac{1}{4}$	30
		$-\frac{1}{4}+1/16$	8
Fines	18	−1/16	18

Bulk samples

Depth below surface (m)	Percentages Fines	Sand	Gravel
*2.2 − 3.3	24	73	3
*3.3 − 4.3	11	35	54
*4.3 − 5.3	6	24	70
*5.3 − 6.3	4	26	70
*6.3 − 7.5	16	52	32
*7.5 − 9.2	37	62	1
*9.2 − 10.2	15	42	43
*10.2 − 11.3	18	31	51

COMPOSITION

Depth below surface (m)	Percentages (in 4.75 to 9.5 mm fractions) Flint	Quartz	Ironstone	Chalk	Limestone
1.2 − 2.2	62	25	0	13	0
2.2 − 3.3	71	8	4	0	17
3.3 − 4.3	80	4	3	4	9
4.3 − 5.3	87	4	2	2	5
5.3 − 6.3	81	5	2	3	9
6.3 − 7.5	83	3	4	4	6
7.5 − 9.2	91	0	0	9	0
9.2 − 10.2	85	4	2	4	5
10.2 − 11.3	78	3	5	2	12
Mean	79	7	2	5	7

SU 78 SE 24	7721 8068	Lower Bolney Farm, Henley	Block C

Surface level (+ 33.5 m) + 110 ft
Water struck at + 31.4 m
Shell and auger (modified) 6 in (152 mm) diameter
August 1972

Overburden 2.1 m (7.0 ft)
Mineral 4.7 m (15.5 ft)
Bedrock 0.5 m+ (1.5 ft+)

LOG

Geological Classification	Lithology	Thickness m	ft	Depth m	ft
	Soil	0.2	(0.5)	0.2	(0.5)
Alluvium	Clay, silty and sandy, dark brown, with scattered flint pebbles	0.8	(2.5)	1.0	(3.5)
	Silt, with sand and trace flint gravel, brown	1.1	(3.5)	2.1	(7.0)
Younger River Gravels (undifferentiated)	'Clayey' gravel Gravel: coarse with fine, subangular to subrounded, with some rounded flint, with chalk below 4.9 m Sand: medium with some coarse and a little fine, subrounded to rounded quartz with some angular to sub-angular flint	4.7	(15.5)	6.8	(22.5)
Upper Chalk	Chalk	0.5+	(1.5+)	7.3	(24.0)

GRADING

Mean for deposit

	%	mm	%
Gravel	58	+16	34
		−16+4	24
Sand	30	−4+1	7
		−1+¼	20
		−¼+1/16	3
Fines	12	−1/16	12

Bulk samples

Depth below surface (m)	Percentages Fines	Sand	Gravel
*2.1 − 3.1	9	23	68
*3.1 − 4.1	13	23	64
*4.1 − 5.1	11	33	56
*5.1 − 6.1	17	41	42
*6.1 − 6.8	10	31	59

SU 78 SE 25 7812 8486 Near Hambleden Lock, Mill End Block C

Surface level (+ 29.9 m) + 98 ft
Water struck at + 28.2 m
Shell and auger (modified) 6 in (152 mm) diameter
August 1972

Overburden 1.7 m (5.5 ft)
Mineral 8.2 m (27.0 ft)
Bedrock 0.5 m+ (1.5 ft+)

LOG

Geological Classification	Lithology	Thickness m	Thickness ft	Depth m	Depth ft
	Soil	0.3	(1.0)	0.3	(1.0)
Alluvium	Clay, silty, light brown, scattered pebbles	0.5	(1.5)	0.8	(2.5)
	Silt, bluish grey	0.9	(3.0)	1.7	(5.5)
Younger River Gravels (undifferentiated)	Gravel Gravel: fine to coarse, subangular to rounded flint with quartz, limestone and chalk, the latter being particularly abundant below 8.8 m, and with some ironstone Sand: coarse and medium, subrounded to rounded quartz with some chalk and with angular to subangular flint, brown	8.2	(27.0)	9.9	(32.5)
Middle Chalk	Chalk	0.5+	(1.5+)	10.4	(34.0)

GRADING

Mean for deposit

	%	mm	%
Gravel	67	+16	36
		−16+4	31
Sand	30	−4+1	15
		$-1+\frac{1}{4}$	14
		$-\frac{1}{4}+1/16$	1
Fines	3	−1/16	3

Bulk samples

Depth below surface (m)	Fines	Sand	Gravel
*1.7 – 2.7	4	37	59
*2.7 – 3.7	1	24	75
*3.7 – 4.7	0	34	66
*4.7 – 5.7	4	23	73
*5.7 – 6.7	7	39	54
*6.7 – 7.7	1	27	72
*7.7 – 8.8	0	24	76
*8.8 – 9.9	4	38	58

COMPOSITION

Percentages (in 4.75 to 9.5 mm fractions)

Depth below surface (m)	Flint	Quartz	Ironstone	Chalk	Limestone
1.7 – 2.7	75	9	5	1	10
2.7 – 3.7	79	6	3	1	11
3.7 – 4.7	81	8	3	1	7
4.7 – 5.7	71	4	6	2	17
5.7 – 6.7	77	5	3	1	11
6.7 – 7.7	82	6	2	5	3
7.7 – 8.8	80	4	3	7	5
8.8 – 9.9	66	1	2	27	3
Mean	76	5	3	6	10

SU 78 SE 26 | 7821 8068 | Near Hennerton House, Wargrave | Block C

Surface level (+ 32.6 m) + 107 ft
Water struck at + 286.6 m
Shell and auger (modified) 6 in (152 mm) diameter
August 1972

Waste 5.6 m (18.5 ft)
Bedrock 0.2 m+ (0.5 ft+)

LOG

Geological Classification	Lithology	Thickness m	Thickness ft	Depth m	Depth ft
Alluvium	Clay, silty, dark brown and bluish green	2.9	(9.5)	2.9	(9.5)
	Clay, silty, brown, with chalk and flint pebbles	1.4	(4.5)	4.3	(14.0)
	Sand, and chalk and flint gravel	1.3	(4.5)	5.6	(18.5)
Upper Chalk	Chalk	0.2+	(0.5+)	5.8	(19.0)

SU 78 SE 27 | 7972 8399 | Near Westfield Farm, Medmenham | Block C

Surface level (+ 30.5 m) + 100 ft
Water struck at + 29.7 m
Shell and auger (modified) 6 in (152 mm) diameter
January 1973

Overburden 0.8 m (2.5 ft)
Mineral 6.1 m (20.0 ft)
Bedrock 0.5 m+ (1.5 ft+)

LOG

Geological Classification	Lithology	Thickness m	ft	Depth m	ft
	Soil	0.3	(1.0)	0.3	(1.0)
Younger River Gravels (undifferentiated)	Clay, silty, mottled light brown and greyish brown, scattered pebbles	0.5	(1.5)	0.8	(2.5)
	Gravel Gravel: fine to coarse with some cobbles, subangular to subrounded flint, with ironstone, chalk and limestone and with some well rounded quartz Sand: medium to coarse, flint and quartz, pale grey	6.1	(20.0)	6.9	(22.5)
Middle Chalk	Chalk	0.5+	(1.5+)	7.4	(24.5)

GRADING

Mean for deposit

	%	mm	%
Gravel	66	+16	30
		−16+4	36
Sand	32	−4+1	15
		$-1+\frac{1}{4}$	16
		$-\frac{1}{4}+1/16$	1
Fines	2	−1/16	2

Bulk samples

Depth below surface (m)	Percentages Fines	Sand	Gravel
*0.8 − 1.8	3	36	61
*1.8 − 2.8	2	17	81
*2.8 − 3.8	2	32	66
*3.8 − 4.8	2	27	71
*4.8 − 5.8	1	31	68
*5.8 − 6.4	2	40	58
*6.4 − 6.9	8	47	45

COMPOSITION

Depth below surface (m)	Percentages (in 4.75 to 9.5 mm fractions) Flint	Quartz	Ironstone	Chalk	Limestone
0.8 − 1.8	No information				
1.8 − 2.8	89	1	2	5	3
2.8 − 3.8	72	2	7	5	14
3.8 − 4.8	No information				
4.8 − 5.8	84	3	4	2	7
5.8 − 6.4	No information				
6.4 − 6.9	45	8	17	18	12
Mean	71	4	8	8	9

SU 78 SE 28 7971 8245 Near Rosehill Wood, east of Remenham Hill

Surface level (+ 97.2 m) + 319 ft
Water not struck
Shell and auger (modified) 6 in (152 mm) diameter

Waste 1.5 m (5.0 ft)
Bedrock 1.0 m+ (3.5 ft+)

LOG

Geological Classification	Lithology	Thickness m	ft	Depth m	ft
	Soil	0.3	(1.0)	0.3	(1.0)
Clay-with-flints	Clay, brown, silty, with large subangular to rounded flints	0.5	(1.5)	0.8	(2.5)
	Clay, light brown and grey, silty, with pebbles of flint and chalk	0.7	(2.5)	1.5	(5.0)
Upper Chalk	Chalk, with flints	1.0+	(3.5+)	2.5	(8.0)

SU 78 SE 29 7936 8162 Near Goulders Cottages, Cockpole Green

Surface level (+ 96.6 m) + 307 ft
Water not struck
Shell and auger (modified) 6 in (152 mm) diameter
January 1973

Waste 1.3 m (4.5 ft)
Bedrock 0.7 m+ (2.5 ft+)

LOG

Geological Classification	Lithology	Thickness m	ft	Depth m	ft
	Soil	0.3	(1.0)	0.3	(1.0)
Clay-with-flints	Clay, brown, with fine to medium pebbles of subangular to subrounded flint and some chalk	0.3	(1.0)	0.6	(2.0)
	Clay, chalky, yellowish brown with abundant fine to medium pebbles of chalk and some flint	0.7	(2.5)	1.3	(4.5)
Upper Chalk	Chalk, with flints	0.7+	(2.5+)	2.0	(6.5)

APPENDIX G: LIST OF WORKINGS

In 1976 the sand and gravel pits listed were known to be operational. The main worked-out areas are shown on the map accompanying the report.

Location	Grid reference	Deposits
Dean's Farm, Caversham	734 744	Younger River Gravels
Sonning Eye	748 755	Younger River Gravels
Colemansmoor Farm	775 725	Younger River Gravels
Black Bridge	781 728	Younger River Gravels
Whistleymill Farm	788 749	Younger River Gravels

APPENDIX H: CONVERSION TABLE, METRES TO FEET (TO NEAREST 0.5 FT).

m	ft	m	ft	m	ft	m	ft	m	ft
0.1	0.5	6.1	20	12.1	39.5	18.1	59.5	24.1	79
0.2	0.5	6.2	20.5	12.2	40	18.2	59.5	24.2	79.5
0.3	1	6.3	20.5	12.3	40.5	18.3	60	24.3	79.5
0.4	1.5	6.4	21	12.4	40.5	18.4	60.5	24.4	80
0.5	1.5	6.5	21.5	12.5	41	18.5	60.5	24.5	80.5
0.6	2	6.6	21.5	12.6	41.5	18.6	61	24.6	80.5
0.7	2.5	6.7	22	12.7	41.5	18.7	61.5	24.7	81
0.8	2.5	6.8	22.5	12.8	42	18.8	61.5	24.8	81.5
0.9	3	6.9	22.5	12.9	42.5	18.9	62	24.9	81.5
1.0	3.5	7.0	23	13.0	42.5	19.0	62.5	25.0	82
1.1	3.5	7.1	23.5	13.1	43	19.1	62.5	25.1	82.5
1.2	4	7.2	23.5	13.2	43.5	19.2	63	25.2	82.5
1.3	4.5	7.3	24	13.3	43.5	19.3	63.5	25.3	83
1.4	4.5	7.4	24.5	13.4	44	19.4	63.5	25.4	83.5
1.5	5	7.5	24.5	13.5	44.5	19.5	64	25.5	83.5
1.6	5	7.6	25	13.6	44.5	19.6	64.5	25.6	84
1.7	5.5	7.7	25.5	13.7	45	19.7	64.5	25.7	84.5
1.8	6	7.8	25.5	13.8	45.5	19.8	65	25.8	84.5
1.9	6	7.9	26	13.9	45.5	19.9	65.5	25.9	85
2.0	6.5	8.0	26	14.0	46	20.0	65.5	26.0	85.5
2.1	7	8.1	26.5	14.1	46.5	20.1	66	26.1	85.5
2.2	7	8.2	27	14.2	46.5	20.2	66.5	26.2	86
2.3	7.5	8.3	27	14.3	47	20.3	66.5	26.3	86.5
2.4	8	8.4	27.5	14.4	47	20.4	67	26.4	86.5
2.5	8	8.5	28	14.5	47.5	20.5	67.5	26.5	87
2.6	8.5	8.6	28	14.6	48	20.6	67.5	26.6	87.5
2.7	9	8.7	28.5	14.7	48	20.7	68	26.7	87.5
2.8	9	8.8	29	14.8	48.5	20.8	68	26.8	88
2.9	9.5	8.9	29	14.9	49	20.9	68.5	26.9	88.5
3.0	10	9.0	29.5	15.0	49	21.0	69	27.0	88.5
3.1	10	9.1	30	15.1	49.5	21.1	69	27.1	89
3.2	10.5	9.2	30	15.2	50	21.2	69.5	27.2	89
3.3	11	9.3	30.5	15.3	50	21.3	70	27.3	89.5
3.4	11	9.4	31	15.4	50.5	21.4	70	27.4	90
3.5	11.5	9.5	31	15.5	51	21.5	70.5	27.5	90
3.6	12	9.6	31.5	15.6	51	21.6	71	27.6	90.5
3.7	12	9.7	32	15.7	51.5	21.7	71	27.7	91
3.8	12.5	9.8	32	15.8	52	21.8	71.5	27.8	91
3.9	13	9.9	32.5	15.9	52	21.9	72	27.9	91.5
4.0	13	10.0	33	16.0	52.5	22.0	72	28.0	92
4.1	13.5	10.1	33	16.1	53	22.1	72.5	28.1	92
4.2	14	10.2	33.5	16.2	53	22.2	73	28.2	92.5
4.3	14	10.3	34	16.3	53.5	22.3	73	28.3	93
4.4	14.5	10.4	34	16.4	54	22.4	73.5	28.4	93
4.5	15	10.5	34.5	16.5	54	22.5	74	28.5	93.5
4.6	15	10.6	35	16.6	54.5	22.6	74	28.6	94
4.7	15.5	10.7	35	16.7	55	22.7	74.5	28.7	94
4.8	15.5	10.8	35.5	16.8	55	22.8	75	28.8	94.5
4.9	16	10.9	36	16.9	55.5	22.9	75	28.9	95
5.0	16.5	11.0	36	17.0	56	23.0	75.5	29.0	95
5.1	17	11.1	36.5	17.1	56	23.1	76	29.1	95.5
5.2	17	11.2	36.5	17.2	56.5	23.2	76	29.2	96
5.3	17.5	11.3	37	17.3	57	23.3	76.5	29.3	96
5.4	17.5	11.4	37.5	17.4	57	23.4	77	29.4	96.5
5.5	18	11.5	37.5	17.5	57.5	23.5	77	29.5	97
5.6	18.5	11.6	38	17.6	57.5	23.6	77.5	29.6	97
5.7	18.5	11.7	38.5	17.7	58	23.7	78	29.7	97.5
5.8	19	11.8	38.5	17.8	58.5	23.8	78	29.8	98
5.9	19.5	11.9	39	17.9	58.5	23.9	78.5	29.9	98
6.0	19.5	12.0	39.5	18.0	59	24.0	78.5	30.0	98.5

REFERENCES

ALLEN, V.T. 1936. Terminology of medium-grained sediments. Rep. Natl. Res. Counc. Washington, 1935-1936. App. 1. Rep. Comm. Sedimentation. pp. 18-47.

ARCHER, A.A. 1969. Background and problems of an assessment of sand and gravel resources in the United Kingdom. Proc. 9th Commonw. Min. Metall. Congr. 1969, Vol. 2, Mining and Petroleum Geology, pp. 495-508. (London: The Institute of Mining and Metallurgy.)

——— 1970a. Standardisation of the size classification of naturally occurring particles. Geotechnique, Vol. 20, pp. 103-107.

——— 1970b. Making the most of metrication. Quarry Managers' J., Vol. 54, No. 6, pp. 223-227.

ATTERBERG, A. 1905. Die rationelle Klassifikation der Sande und Kiese. [The rational classification of sands and gravels.] Chem. Z., Vol. 29, pp. 195-198.

BLAKE, J. H. 1903. The geology of the country around Reading. Mem. Geol. Surv. G.B.

BRITISH STANDARD 1377. 1967. Methods of testing soils for civil engineering purposes. (London: British Standards Institution) 233 pp.

BUREAU OF MINES AND GEOLOGICAL SURVEY. 1948. Mineral Resources of the United States (Washington, D.C.: Public Affairs Press), pp. 14-17.

HARRIS, P.M., THURRELL, R.G., HEALING, R.A. and ARCHER, A.A. 1974. Aggregates in Britain. Proc. R. Soc., Ser. A, Vol. 339, pp. 329-353.

JUKES-BROWNE, A.J. and WHITE, H.J. 1908. The geology of the country around Henley-on-Thames and Wallingford. Mem. Geol. Surv. G.B.

LANE, E.W. and others. 1947. Report of the subcommittee on sediment terminology. Trans. Am. Geophys. Union, Vol. 28, pp. 936-938.

OSBORNE WHITE, H.J. 1907. The geology of the country around Hungerford and Newbury. Mem. Geol. Surv. G.B.

PETTIJOHN, F.J. 1957. Sedimentary rocks. 2nd ed. (London: Harper and Row.)

THURRELL, R.G. 1971. The assessment of mineral resources with particular reference to sand and gravel. Quarry Managers' J., Vol. 55, pp. 19-25.

TWENHOFEL, W.H. 1937. Terminology of the fine-grained mechanical sediments. Rep. Natl. Res. Coun. Washington, 1936-1937, App. 1, Rep. Comm. Sedimentation, pp. 81-104.

UDDEN, J.A. 1914. Mechanical composition of clastic sediments. Bull. Geol. Soc. Am., Vol. 25, pp. 655-744.

WENTWORTH, C.K. 1922. A scale of grade and class terms for clastic sediments. J. Geol. Vol. 30, pp. 377-392.

——— 1935. The terminology of coarse sediments. Bull. Natl. Res. Coun. Washington, No. 98, pp. 225-246.

WILLMAN, H.B. 1942. Geology and mineral resources of Marseilles, Ottawa and Streator quadrangles. Bull. Illinois State Geol. Surv. 66, pp. 343-344.

Dd 587333 K8

Printed in England for Her Majesty's Stationery Office by Commercial Colour Press, London